T0316738

Systematic Conservation Planning

Systematic Conservation Planning provides a clear, comprehensive guide to the process of deriving conservation area networks that will represent biodiversity in cost-effective ways. The measurement of biodiversity, the design of field collections to sample biodiversity and data treatment methods together provide baseline information on the distribution patterns of biodiversity. The book then describes methods for identifying biodiversity priority areas underpinned by the concept of complementarity. It goes on to describe how biodiversity targets can be set and how multi-criteria analyses can incorporate costs and other constraints into the area selection process, enabling trade-offs between production and protection to be made. The book finishes with a series of case studies and a conclusion that emphasizes the socio-economic and cultural contexts in which conservation planning takes place.

A clear procedure is provided for identifying biodiversity priority areas, underpinned by current best science practice and sound common sense, making this volume of interest not only to graduate students and academic conservation biologists, but also planners and decision-makers dealing with natural resource management together with conservation NGOs.

CHRIS MARGULES is Executive Director at the Melanesia Centre for Biodiversity Conservation, Conservation International.

SAHOTRA SARKAR is a Professor in the Section of Integrative Biology at the University of Texas at Austin.

ECOLOGY, BIODIVERSITY AND CONSERVATION

The world's biological diversity faces unprecedented threats. The urgent challenge facing the concerned biologist is to understand ecological processes well enough to maintain their functioning in the face of the pressures resulting from human population growth. Those concerned with the conservation of biodiversity and with restoration also need to be acquainted with the political, social, historical, economic and legal frameworks within which ecological and conservation practice must be developed. This series will present balanced, comprehensive, up-to-date and critical reviews of selected topics within the sciences of ecology and conservation biology, both botanical and zoological, and both "pure" and "applied". It is aimed at advanced final-year undergraduates, graduate students, researchers and university teachers, as well as ecologists and conservationists in industry, government and the voluntary sectors. The series encompasses a wide range of approaches and scales (spatial, temporal, and taxonomic), including quantitative, theoretical, population, community, ecosystem, landscape, historical, experimental, behavioral and evolutionary studies. The emphasis is on science related to the real world of plants and animals, rather than on purely theoretical abstractions and mathematical models. Books in this series will, wherever possible, consider issues from a broad perspective. Some books will challenge existing paradigms and present new ecological concepts, empirical or theoretical models, and testable hypotheses. Other books will explore new approaches and present syntheses on topics of ecological importance.

Systematic Conservation Planning

CHRISTOPHER R. MARGULES

Conservation International and CSIRO Sustainable Ecosystems,
Tropical Forest Research Centre, Atherton, Queensland

SAHOTRA SARKAR

Section of Integrative Biology, and Department
of Philosophy, University of Texas at Austin

CAMBRIDGE
UNIVERSITY PRESS

CAMBRIDGE
UNIVERSITY PRESS

Shaftesbury Road, Cambridge CB2 8EA, United Kingdom

One Liberty Plaza, 20th Floor, New York, NY 10006, USA

477 Williamstown Road, Port Melbourne, VIC 3207, Australia

314–321, 3rd Floor, Plot 3, Splendor Forum, Jasola District Centre, New Delhi – 110025, India

103 Penang Road, #05–06/07, Visioncrest Commercial, Singapore 238467

Cambridge University Press is part of Cambridge University Press & Assessment, a department of the University of Cambridge.

We share the University's mission to contribute to society through the pursuit of education, learning and research at the highest international levels of excellence.

www.cambridge.org
Information on this title: www.cambridge.org/9780521703444

First published 2007

A catalogue record for this publication is available from the British Library

ISBN   978-0-521-87875-3   Hardback
ISBN   978-0-521-70344-4   Paperback

Contents

The colour plates are situated between pages 130 and 131.

Acknowledgments

The collaboration from which this book emerged began at the *Wissenschaftskolleg zu Berlin* where both authors were fellows in the mid-1990s. We thank the *Kolleg* for its support. We also thank Paul Griffiths of the University of Queensland Biohumanities Project for providing us with hospitality while we finished the book. A huge thanks also to Elva Castino for all her help with finalizing material and getting everything ready for publication.

The text of the first two parts of Chapter 3 draws heavily on material from Margules *et al.* (2002) and Williams *et al.* (2002). The text of Chapter 7 draws heavily on Moffett & Sarkar (2005) and Moffett *et al.* (2006). We thank Jim Dyer, David Hilbert, Alex Moffett, Bob Pressey and Paul Williams for permission to use this material, as well as for many comments and discussions over the years.

For commenting on the entire manuscript we thank Trevon Fuller, Justin Garson, David Grice, James Justus and Liz Poon. For commenting on individual chapters we thank Mark Burgman, Dan Faith, Simon Ferrier, Nick Nicholls and Michael Usher.

For critical comment and discussion on many of the ideas in this book, sometimes over many years, we would also like to thank Anshu Aggarwal, Mike Austin, Mark Boyce, Helen Cortes-Burns, Chris Humphries, Patricia Illoldi, Chris Johnson, Chris Kelley, Vanessa Lujan, Camille Parmesan, Samraat Pawar, Hugh Possingham, Karen Richardson, Victor Sánchez-Cordero, Helen Sarakinos, Mike Singer, Dick Vane-Wright, Paul Walker, Kristen Williams and Paul Williams.

CRM would like to record a very special thanks to Janice Marion Margules, 1955–94, for her tireless support and encouragement of the ideas embodied in this book.

1 · Introduction

The earth is experiencing a wave of extinctions possibly unprecedented in recent evolutionary history. The ecological advantage conferred upon humans by the evolution of reason and language has enabled us to obtain competitive dominance over all other species and expand our realized niche to include components of almost all habitats on the planet. We have simplified complex and diverse natural systems and processes for the production of food and shelter, leading to massive extinctions of species (Taylor 2004). The previous six mass extinctions caused by cataclysmic events decimated life on the planet and the current wave shows signs of approaching this level. These previous waves of extinction occurred over millennia; brief in geological time but long in ecological time. Life forms had the opportunity to adapt and evolve. Many species became extinct, but many others prospered and new species evolved. The current wave is not yet 300 years old; a brief aberration even in ecological time. Though humans have undoubtedly changed ecological processes and caused extinctions for longer than that, it was not at the current scale. Perhaps what is most troubling about the current wave of extinctions is that, unlike previous events, it is taking place because of a type of irreversible transformation of the land and even the oceans that is likely to make them unsuitable as habitat for many forms of life into the distant future.

The task for conservation biology is to halt the current extinction wave and chart a course for a future which includes biological diversity not only for its direct contribution to human welfare as a resource, but also because it appeals to important human values (Norton 1987; Sarkar 2005). In that future, nature conservation must exist alongside nature exploitation, not instead of it. Better still, the protection of biodiversity should become integrated into natural resource management that has sustainable livelihoods as an equally important goal. The protection of biodiversity will not win out in direct competition with other needs and aspirations of people (Rosenzweig 2003). Indeed, the research frontiers in conservation biology clearly must link a biophysical understanding of the world around us with a socio-economic understanding of what drives decisions on natural resource use, as well as an understanding of natural resource governance mechanisms and the institutional arrangements that societies use to regulate the exploitation of natural resources. This task of planning for the future is made even more complex by global environmental changes, especially climate change, the detailed consequences of which we can only dimly foresee, and even that only for a very few species. We will return to this issue of linking socio-economic goals with biodiversity conservation in Chapters 5 through 9, especially Chapters 7, 8 and 9. The focus in Chapters 2, 3 and 4 is on methods for deriving biodiversity data sets suitable for use in conservation planning.

1.1 Conservation area networks

Central to the conservation of biological diversity is the establishment of networks of conservation areas which are managed to minimize the risk of extinction. These are often called protected areas, but we use the term conservation areas to include all areas that perform a conservation function, whether they are strictly protected or not (Sarkar 2003). These are the priority areas for the allocation of scarce biodiversity management resources. *Systematic conservation planning* consists of the use of specific protocols to identify such priority areas and separate them from processes which threaten their persistence. Identifying and securing biodiversity priority areas will not in itself protect biodiversity, but networks of priority areas in each region of the world should form the framework upon which other conservation actions build.

The key concept underpinning systematic conservation planning is complementarity. Complementarity is a measure of the contribution an area in a planning region makes to the full complement of biodiversity features: species, assemblages, ecological processes, etc. As discussed below and in later chapters, we can never know the full complement of biodiversity features; consequently we must use partial measures or surrogates and we must set goals for the representation of these surrogates. Complementarity can then be measured by the contribution an area makes to the conservation goal. For example, pretend that the conservation goal is to represent at least one population of all of the vascular plant species that occur in a planning region in a minimal set of biodiversity priority areas. The first area we choose might be the one with the most species. The next area we choose will be the one with the most species that were not represented in the first area and the next area after that will be the one with the most remaining unrepresented species, and so on until all species are represented. There are two important points to note about this process. The first is that as areas are added to the set, the contribution that remaining areas make to the goal changes. This is because some of the species in those areas may have already been contributed by other areas previously selected for the set. The second is that areas with the highest complementarity will not necessarily be those with the most species. In this example, complementarity is measured at each step as the number of unrepresented species. If an area has few species, but they do not occur widely in the planning region, it may have higher complementarity than an area with many species that are widespread throughout the planning region (depending on which species occur in areas that have already been selected). Thus, species richness cannot be used to measure complementarity. There are many variations on the simple area selection process outlined above and many of these are described later in the book. As it turns out, for example, it is often more economical to begin the area-selection process by selecting that area with the rarest species on it and then the one with the highest number of rarest unrepresented species, and so on until all species are represented (Sarkar *et al.* 2002). Vane-Wright *et al.* (1991) coined the term "complementarity," but it appears to have been independently discovered at least four times (Justus & Sarkar 2002), twice in Australia (Kirkpatrick, 1983; Margules & Nicholls 1987; Margules *et al.* 1988), in the United Kingdom (Ackery & Vane-Wright 1984) and in South Africa (Rebelo & Siegfried 1990). Complementarity is described in detail in Chapter 5 (Section 5.4).

Systematic conservation planning is usually implemented with software tools using digital georeferenced data sets and area selection algorithms. Algorithmic approaches ensure that planning exercises are standardized and repeatable – this is the most important sense in which conservation planning is systematic. However, this numerical computer-based approach has led to some criticisms, in particular that large data sets are needed to run the software (Redford *et al.* 1997; Prendergast *et al.* 1999), that local expertise is not consulted and incorporated (Redford *et al.* 1997) and that they are expensive and money could better be spent acquiring reserves (Prendergast *et al.* 1999). Any reading of the conservation planning literature will show that these criticisms are unjustified. There is no substitute for local expertise, which should always be consulted. In fact, systematic conservation planning works best when local experts are the ones doing the planning. The methods are inexpensive, can be used without computers and with minimal data. It is just that computers speed up the process and enable large data sets to be used efficiently when they are available. Also, the recent growth in biological and environmental databases, in part due to rapid improvements in remote sensing and modeling, means that there is now almost no terrestrial region on Earth with so little data that systematic planning tools cannot improve policy formulation.

Where population density is high and land is in private ownership, conservation agencies have, in the past, found it necessary to take opportunities whenever and wherever they occurred believing that when land is privately owned there is no opportunity for a more strategic systematic approach. However, if conservation NGOs (non-government organizations) such as The Nature Conservancy in the United States and Bush Heritage in Australia are purchasing land, as they currently are, they need to know what the genuine biodiversity priorities are or they will not make best use of their funds. If they purchase the cheapest land or just whatever becomes available they will not make the contributions to biodiversity conservation that they could make, and which are sorely needed. As several past studies have shown, ad hoc conservation-area selection has been remarkably cost-ineffective (Pressey *et al.* 1996; Pressey & Cowling 2001). This is because conservation areas all over the world contain a biased sample of biodiversity from ecosystems and habitats that were selected because they are remote and inaccessible, or they are unfit for alternative uses such as agriculture, which means they are cheap. The identification of biodiversity priority areas should be based on good science and sound common sense and not on such ad hoc grounds. Some priority areas will inevitably be expensive. But even if they prove too expensive to buy, we should know about them. They might become subject to policy change and/or management agreements in future. Planning tools support that process. They do not replace it. Conservation planning is a dynamic iterative process and these planning tools are designed for decision support to help local experts identify good policy options, not formulate policy on their behalf.

In the United States, Texas provides a striking example of the potential continued inefficient use of resources when conservation policy is not determined by systematic planning (Alford 2005). In 2003 tax exemptions given to landowners for maintaining or restoring their land as wildlife habitat were estimated to be over $1 billion. Land management options included habitat control (such as clearing juniper overgrowth), erosion control, control of predators (such as feral cats or dogs), providing water resources, providing supplemental food for wild animals, growing nutritional plants, providing animal shelters (such as leaving dead trees for bird nests) and conducting animal censuses. There is no coordination among efforts of individual landowners. Although those who receive the tax break have to formulate explicit wildlife management plans, these do not have to be filed with any state agency, but only be appraised by county officials who need not have any wildlife expertise. There is no monitoring to determine whether plans are successful in promoting wildlife. In some cases anecdotal evidence suggests no change: the deer population growth rate remained the same before and after the policy was introduced in 1995. In other cases, there may be some success: while the decline of wild turkey populations was 54% in the five years preceding 1995, it was only 9% in the subsequent five years. Given the lack of coordination and monitoring it is impossible to evaluate what is being conserved or restored and how effectively. Perhaps more importantly, there is no prioritization of land on the basis of its potential contribution to a biodiversity goal, given appropriate management, and therefore no prioritization of the landowners who should receive these tax breaks. This is the type of problem that systematic conservation planning aims to mitigate.

In some places where land is privately owned and there are few opportunities for biodiversity protection using traditional approaches, market-based instruments are being tested and evaluated. In a recent pilot study in the highly fragmented landscape of the Western Australian wheatbelt called "Auction for Landscape Recovery" (Gole *et al.* 2005), landowners were invited to nominate portions of their holdings and propose management actions for these portions and then bid for funds to implement those actions. Bids were judged on two criteria. One was the contribution the nominated areas made to a regional biodiversity goal, i.e., complementarity, and the other was the likely success of the proposed management actions. This is an example of how the systematic approach is beginning to find wide application in the development and implementation of conservation policies and practices.

1.2   What do we mean by biodiversity?

Biodiversity consists of the biological variety bequeathed to us by evolutionary processes over millennia. It is what we have to conserve if we do not want to squander this inheritance. The biological realm is characterized by variability and complexity at every level of structural, taxonomic and functional organization. The term "biodiversity" was introduced in the mid 1980s as a contraction of "biological diversity" to refer to the totality of this variability (Takacs 1996). From a biological perspective, all such diversity is important because it provides the raw material for evolution. Ideally, all of it should be conserved.

Ecosystem processes such as nutrient cycling, the movement of water and energy and the dispersal of propagules are necessary for the persistence of biodiversity, but in this book we will restrict the definition of "biodiversity" to biological patterns such as the distribution patterns of biological entities and dynamic behavioral phenomena such as migrations. Two hierarchical schemes are used for the classification of biological entities (Sarkar 1998). One is a spatial (or ecological) hierarchy starting with molecules and macromolecules, then cell organelles, cells, individuals, populations and metapopulations, communities, ecosystems and ultimately the biosphere. The second is a taxonomic hierarchy from alleles to loci, linkage groups, genotypes, subspecies, species, genera, families, orders, classes, phyla and kingdoms. Both hierarchies reflect evolutionary history and are constrained by evolutionary mechanisms. Since the future for biodiversity is dependent on evolutionary processes, understanding the relationships between these two hierarchies and phylogeny is also necessary for the formulation of successful conservation strategies. In practice we never know these relationships fully but must proceed anyway, accepting that conservation priorities and strategies will change as knowledge accumulates.

There are two points to note about both hierarchies. The first is that they are not clean and tidy in the sense that biological entities fall neatly into well-defined classes without exception. The second is that classes at all levels in each hierarchy are heterogeneous: there is variety within each class at every level. The variety of viable biological configurations at all hierarchical levels is extremely large, currently unknown, and probably unmeasurable. This second point is almost trivial, but emphasizes the complexity of biodiversity. Almost any two populations, even of the same subspecies, differ in their genetic profiles. Except for some clonal organisms, almost any two individuals of the same species have different genotypes. There are virtually no two identical ecological communities, and so on. The first point is equally important when we consider how to describe biodiversity. While some entities such as organelles and cells are reasonably well defined, examples such as fungi, symbionts and clonal organisms show that even biological "individuals" are not always precisely defined. Asexual species are notoriously difficult to define and even sexual species, usually defined by the ability to interbreed and produce offspring (Mayr 1957), cause problems, for instance in the case of so-called "ring species." In such species, "rings" are composed of circular geographically successive populations that evolved from a single ancestral population. Each pair of neighboring populations can interbreed except for the two terminal ones which, by definition, should now be regarded as different species. Examples include the greenish warbler, *Phylloscopus trochiloides*, populations surrounding Tibet and the salamander, *Ensatina eschscholtzii*, surrounding the central valley of California (Wake 2001).

As we will discuss in detail in Chapter 2, biodiversity, described in such a complex way, is impossible to estimate or quantify in the field; it is not an operational concept (Sarkar 2002; Sarkar & Margules 2002). In practice, the concept of biodiversity has often been simplified to refer to diversity at three levels of organization: genes (alleles), species and ecosystems (e.g., Meffe & Carroll 1994; Vermeulen & Koziell 2002 and see Lindenmayer & Burgman 2005 for a detailed discussion of biodiversity at different organizational and functional levels). Any such definition is necessarily partly conventional. However, that does not mean that it must be arbitrary. Such a definition is conventional because we know that it does not include all of what we mean by diversity. For instance, the definition of biodiversity as diversity of alleles, species and ecosystems excludes interspecific hybrids. More importantly, it excludes biological phenomena such as the annual migrations of north American monarch butterflies (*Danaus plexippus*), which may be endangered and deserving protection (Brower & Malcolm 1991; see Chapter 2). Nevertheless, the definition is not arbitrary because focusing conservation efforts on genes, species and ecosystems will protect much of the diversity within species, taxonomic diversity at levels higher than species and many communities (see Chapter 2).

As Austin and Margules (1986) first explicitly pointed out, the concept of biodiversity must be operationalized through the use of "surrogates," features of the landscape such as the presence of species or other taxa, habitat type, etc., that can *in principle* be quantified and assessed in the field. Surrogates that are used to represent biodiversity in its full generality we call "true" surrogates. For example, "biodiversity in general" is often thought to be adequately represented by species diversity. The identification and adoption of true surrogates operationalizes the concept of biodiversity for systematic conservation planning. But, as we noted earlier, any choice of a true surrogate set is partly conventional. Moreover, since general biodiversity cannot be quantified, the success of a true surrogate set in representing general biodiversity also cannot be quantitatively assessed. What we must do is give good reasons for what we choose to be the true surrogate set in a given planning context. Quite often, we are not in a position even to assess the full distribution of a true surrogate set. For instance, if we take all species within even well-documented taxa such as mammals and birds, there are many areas of the world for which we will not have all the distributional data we need (see Chapter 3). In many situations we may not even be able to model such distributions completely (see, however, Chapter 4). We then have to use "estimator surrogates" for biodiversity; those features which we can *in practice* quantify and assess in the field. The question whether an estimator surrogate set adequately represents a true surrogate set is amenable to quantitative assessment, as we discuss in detail in Chapter 2.

1.3 Systematic conservation planning

Systematic conservation planning is a structured step-wise approach to mapping conservation area networks, with feedback, revision and re-iteration, where needed, at any stage. Though prioritizing new areas for conservation is central to systematic conservation planning, this process does not ignore or throw away, literally or metaphorically, existing conservation areas or networks. In almost all regions of the world, there is a heritage of conservation areas, which more than likely have been accumulated opportunistically and are therefore unrepresentative of regional biodiversity. Systematic planning normally accepts these imperfect networks and maps onto, or builds on, what already exists, with the objective of transforming them into better networks. In addition, analyses of the extent to which existing conservation areas contribute to regional biodiversity goals might provide options for future rationalization. For example, it may be possible to trade existing conservation areas making low contributions to biodiversity representation for new areas that would make higher contributions.

Margules and Pressey (2000), Groves *et al.* (2002), Cowling and Pressey (2003) and Sarkar (2004) identified several distinguishing characteristics of systematic conservation planning. In the first place, it requires the identification and engagement of stakeholders. The people who rely on or influence the use of natural resources in the planning region must be party to the planning process or it will fail. This point deserves emphasis: far too often conservation policies have failed because people who have a stake in the land or water that is the subject of those policies have not been consulted (Sarkar 1999; Justus & Sarkar 2002). Sometimes, when the planners have come from international organizations or distant national governments, this has led to accusations of paternalism, colonialism and other ideological remnants of the Northern colonial era (Guha 1989). Recent reports have documented the creation of "conservation refugees," people involuntarily displaced by conservation policies such as the creation of reserves and the abrogation of traditional resource-use rights such as hunting (Dowie 2005). Throughout this book we emphasize the point that systematic conservation planning must be viewed as part of social policy which explicitly recognizes and addresses these issues.

Systematic conservation planning also requires that clear choices are made about how biodiversity is to be measured and mapped. This is partly the surrogacy issue raised in Section 1.2 above and discussed in detail in Chapter 2. But it also includes collecting data, building and managing data sets and databases, and carrying out data treatments to derive the chosen surrogates. These issues are discussed in Chapters 3 and 4. Similarly, clear goals have to be set and preferably then translated into explicit measurable conservation targets. Next, the extent to which conservation goals have been met in existing reserves or priority areas must be recognized so that the simple explicit methods described in Chapter 5 can be used to identify new priority areas to complement existing conservation areas in achieving the set goals. Finally, explicit criteria for implementing planning choices on the ground have to be formulated and applied, objectives for individual conservation areas have to be set, the achievement of those objectives has to be monitored and appropriate management actions have to be taken to ensure continued contribution of individual conservation areas to overall conservation goals. This book addresses all of these issues, though Chapter 6 is the only one to focus on the last, while also addressing issues of threat and vulnerability at all stages in the process.

Table 1.1, modified from Margules and Pressey (2000) and Sarkar (2004), describes this overall approach in eleven more-detailed stages. The first stage is stakeholder engagement. Stakeholders will often be local residents, farmers or pastoralists, but can also include government agencies responsible for managing natural resources such as water and forests, non-government organizations (NGOs), including conservation NGOs, both local and global, and industries, e.g., mining and agri-businesses. Stakeholders include all those people who have decision-making powers over a region, all those who will be affected by the conservation plans that are formulated, those with scientific or other expertise about the region and those who may commit resources for conservation planning and implementation. For a conservation plan to be successful, the involvement of

Table 1.1 *Systematic conservation planning*

---

(1) Identify stakeholders for the planning region:
- Stakeholders include: (a) those who have decision-making powers; (b) those who will be affected by conservation plans for the region; (c) those with expertise about the region and (d) those who may commit resources for conservation plans;
- Include both local and global stakeholders;
- Ensure transparency in the involvement of all stakeholders from the beginning.

(2) Compile, assess, and refine biodiversity and socio-economic data for the region:
- Compile available geographical distribution data on as many biotic and environmental parameters as possible at every level of organization;
- Compile available socio-economic data, including values for alternate uses, resource ownership and infrastructure;
- Collect relevant new data to the extent feasible within available time; remote-sensing data should be easily accessible; systematic surveys at the level of species (or lower levels) will rarely be possible;
- Assess conservation status for biotic entities, for instance, their rarity, endemism and endangerment;
- Assess the reliability of the data, formally and informally; in particular, critically analyze the process of data selection;
- When data do not reflect representative samples of the landscape, correct for bias and model distributions.

(3) Identify biodiversity surrogates for the region:
- Choose true surrogate sets for biodiversity (representing general "biodiversity") for part of the region; be explicit about criteria used for this choice;
- Choose alternate estimator surrogate sets (for representing true surrogate sets in the planning process);
- Prioritize sites using true surrogate sets; prioritize sites using as many combinations of estimator surrogate sets as feasible and compare them;
- Potentially also use other methods of surrogacy analysis to assess estimator surrogate sets, including measures of spatial congruence between plans formulated using the true and estimator surrogate sets;
- Assess which estimator surrogate set is best on the basis of (a) economy and (b) representation.

(4) Establish conservation targets and goals:
- Set quantitative targets for surrogate coverage;
- Set quantitative targets for total network area;
- Set quantitative targets for minimum size for population, unit area, etc.;
- Set design criteria such as shape, size, dispersion, connectivity, alignment and replication;
- Set precise goals for criteria other than biodiversity, including socio-political criteria.

Table 1.1 (*cont.*)

(5) Review the existing conservation-area network (CAN):
- Estimate the extent to which conservation targets and goals are met by the existing set of conservation areas;
- Determine the prognosis for the existing CAN;
- Refine the first estimate.

(6) Prioritize new areas for potential conservation action:
- Using principles such as complementarity, rarity and endemism, prioritize areas for their biodiversity content to create a set of potential conservation-area networks;
- Starting with the existing CAN, repeat the process of prioritization to compare results;
- Incorporate socio-political criteria, such as various costs, if desired, using a trade-off analysis;
- Incorporate design criteria such as shape, size, dispersion, connectivity, alignment and replication, if desired, using a trade-off analysis.
- Alternatively, carry out the last three steps using optimal algorithms.

(7) Assess prognosis for biodiversity within each newly selected area:
- Assess the likelihood of persistence of all biodiversity surrogates in all selected areas. This may include population viability analysis for as many species using as many models as feasible;
- Perform the best feasible habitat-based viability analysis to obtain a general assessment of the prognosis for all species in a potential conservation area;
- Assess vulnerability of a potential conservation area from external threats, using techniques such as risk analysis.

(8) Refine networks of areas selected for conservation action:
- Delete the presence of surrogates from potential conservation areas if the viability of that surrogate is not sufficiently high;
- Run the prioritization protocol again to prioritize potential conservation areas by biodiversity value;
- Incorporate design criteria such as shape, size, dispersion, connectivity, alignment and replication.

(9) Examine feasibility using multi-criteria analysis:
- Order each set of potential conservation areas by each of the criteria other than those used in Stage 6;
- Find all best solutions; discard all other solutions;
- Select one of the best solutions.

(10) Implement a conservation plan:
- Decide on most appropriate legal mode of protection for each targeted place;
- Decide on most appropriate mode of management for persistence of each targeted surrogate;
- If implementation is impossible return to Stage 5;
- Decide on a timeframe for implementation, depending on available resources.

(11) Periodically reassess the network:
- Set management goals in an appropriate timeframe for each protected area;
- Decide on indicators that will show whether goals are met;
- Periodically measure these indicators;
- Return to Stage 1.

stakeholders should be transparent. Identifying and involving stakeholders can be a difficult and laborious process, but, if it is done properly, it can help mitigate threats to potential priority areas and improve the chances that conservation plans will be implemented (Wilson *et al.* 2005; see also Chapter 5, Example 5.5). Some recent planning exercises have emphasized that protocols and tools developed to aid planning need to be easy to use and transparent to stakeholders (Pierce *et al.* 2005; Knight *et al.* 2006; Rouget *et al.* 2006).

The second stage is data collection and treatment. This involves collating existing data, collecting new data if required, and any treatment of data that might be needed for subsequent use in conservation planning. The care and attention given this stage has a major bearing on the quality of the outcome. It can be time-consuming, labor intensive, and scientifically and technically challenging. However, the collection and treatment of the biological and environmental data are crucial components of systematic conservation planning. They can place severe constraints on the planning process if not done properly. We regard the collection and treatment of the biological and environmental data as so important that we have devoted two whole chapters to it (Chapters 3 and 4). Cost and the urgency to act tend to foster the use of existing data held, for example, in museums and herbariums or data that can be derived remotely, for example environmental data such as climate surfaces and other maps. All possible use should be made of such data. However, much greater attention than has been paid in the past should be devoted to the design of surveys to collect new biological records from the field (Margules & Austin 1994; Haila & Margules 1996). Field collections of species records are always being made by museums, herbariums, management agencies and others. Environmental stratification combined with recording the absence as well as the presence of species will deliver data sets that are comprehensive and consistent in detail across entire planning regions (Chapter 3). At this stage it is also desirable to compile as much social and economic data as possible, which may then be used in a trade-off analysis in Stage 6, or alternatively in a multi-criteria analysis at Stage 9 in Table 1.1. Socio-economic data include the expected monetary value of the natural resources in candidate conservation areas, or alternative measures such as timber volume or agricultural potential (as in Chapter 8, Section 8.5). They can also include human population density, cultural practices and preferences, including information on land ownership and tenure, as well as infrastructure.

The third stage is to choose biodiversity surrogates. This means selecting those features that are going to be used to represent biodiversity in the planning process. Taxa subsets, species assemblages, and environmental variables and classes, or combinations of two or more of these three, have all been used in conservation planning. Species or other features at risk, and rare or endemic species or features are obvious candidate surrogates. Note that social, cultural and economic significance have at times been just as important as biological significance in the choice of surrogates. Charismatic or iconic species and species with commercial value have often been used as biodiversity surrogates. The choice of surrogates is also a crucial stage and is discussed at length in Chapter 2. The choice of surrogates will always be constrained to a certain extent by what data are available, or realistically obtainable in an acceptable timeframe. Therefore, this choice is never independent of the results of the previous stage.

The fourth stage is to establish planning region goals and targets. The overall goals, identifying priority areas and separating them from threats, were identified above. At this stage, explicit targets for the representation of surrogates within a conservation-area network for the region of interest must be set. Without these targets it is impossible to determine the success or failure of a plan. Typical targets might be populations of a certain size, or a specified number of populations of species, or the spatial extent (percentage of coverage) of assemblages or environmental classes. Again, the actual numbers used here are often not determined, sometimes not even strongly suggested, by biological criteria such as models or empirical data. Viable population sizes are known for only a handful of species in a few habitats. Actual targets most often represent conventions arrived at by biological intuition, or a limited budget. Soulé and Sanjayan (1998) have argued that the achievement of such targets could provide an unwarranted sense of security, suggesting that biodiversity was being protected when in fact the targets have little or no biological meaning. This is undoubtedly true, but, as noted above, conservation areas are priority areas for the allocation of scarce management resources, and their identification is not supposed to provide a complete solution to the problem of protecting biodiversity. Priority areas alone will not protect biodiversity. Identifying priority areas is only one stage in the challenging, but necessary, task of learning to manage whole regions so that ecosystem processes and the biodiversity they give rise to can be sustained along with the generation of livelihoods.

At this stage it is also appropriate to introduce design criteria. Though not usually included in setting targets, these are ecological characteristics of the actual conservation areas – size, shape, dispersion, connectivity, alignment and replication, for example. In meeting conservation goals there seems no doubt that big is better than small, but ecology does not say how big is big enough. In addition, the roles of shape, dispersion, connectivity, alignment and replication in conservation-area networks remain controversial (Diamond 1975; Margules *et al.* 1982; Margules & Pressey 2000). Biogeographical theory, successional pathways, space requirements (especially for wide-ranging species), source–sink population structures, and habitat modification all impact conservation area design (Chapter 6).

The fifth stage is a review of any existing conservation areas within the planning region. The purpose is to determine the extent to which conservation targets have already been met and therefore to identify gaps that need to be filled by new conservation areas. Usually, existing conservation areas have been established on land that has (or had at the time of establishment) little economic value. Gaps in existing conservation-area networks therefore are often in productive areas or close to population centers where competition for natural resources is highest. This emphasizes the need for flexibility in planning and signals the importance of cost trade-offs (Chapters 7 and 8). A careful assessment of the performance of existing conservation areas is critical because, in practice, conservation plans will typically consist of augmenting an existing network rather than creating one from scratch.

The sixth stage involves prioritizing new areas for conservation action to satisfy the targets and goals set in the fourth stage. This stage corresponds to what has previously been thought of as reserve network selection. It is discussed in detail in Chapter 5 and examples of how it has been done are worked through in Chapter 8. The change of terminology recognizes the fact that designating reserves that exclude human habitation is only one of many possible conservation measures that could be implemented in biodiversity priority areas (Sarkar 2003). This stage is at the heart of systematic conservation planning. It implements *complementarity* as a measure of conservation value: one site has greater complementarity than another if it has more biodiversity features (species, assemblages, habitat types, etc.) that have not already met their representation target in the conservation area network. Some measures of complementarity also implement cost trade-offs (Faith & Walker 2002; Faith *et al.* 2001a). This takes advantage of the fact that there are usually many spatial arrangements of selected areas in planning regions that each achieves the conservation goals. A set of conservation priority areas can be sought that optimizes opportunity costs such as agriculture, logging, recreation, industrial development and urbanization, but nevertheless achieves the conservation goal. Stage 9 below represents an alternative approach to taking costs and competing uses of biodiversity into account. In many practical applications, area selection will be limited to those areas that are not obviously irrelevant to conservation because of extensive habitat degradation, for instance, completely built-up areas. Thus, Stage 6 may include a preliminary exclusion of such areas.

The seventh stage assesses the risks to the persistence of biodiversity in selected areas. Threats can come from outside or within. Size, shape, dispersal, connectivity, alignment and replication are some ecological criteria for risk assessment. Suitability for competing uses such as agriculture or urban development increases the probability that a site will be lost to these land uses (Pressey & Taffs 2001). Risk assessment is a difficult task and more remains uncertain than what is known. Chapter 6 considers how vulnerability and threat might be taken into account in conservation planning. Once risks to persistence have been assessed, it is likely that some areas with a poor prognosis will be dropped and prioritization repeated without those areas as candidates. It may be that there are no substitutes for an area with a poor prognosis and in that case a decision has to be made whether to spend potentially scarce management resources on intensive management actions to improve the prognosis or relinquish the biodiversity features of that site to their fate and divert management resources to areas and features with a greater likelihood of long-term persistence. The eighth stage then, is the reiteration of the prioritization process in Stage 6.

The ninth stage attempts to take account of competing uses of land other than biodiversity conservation such as agriculture, recreation, etc. Typically, a number of sets of selected areas (or "solutions" from an area prioritization algorithm) are first produced with each satisfying the biodiversity representation targets. Stakeholders decide the relative importance of different potential uses of land, and these preferences are used to order each of the alternative solutions by all the criteria other than biodiversity. Biological criteria other than representation of surrogates, such as size, shape, dispersal, connectivity, alignment and replication, can also be incorporated in this way through multi-criteria analysis. The best solutions become candidates for implementation and the others are discarded. There are a variety of techniques for carrying out such a multi-criteria analysis, mainly developed by economists and the decision-theory community and only lately being explicitly incorporated into conservation planning (Moffett & Sarkar 2005). The most useful techniques of multi-criteria analysis are discussed in Chapter 7. If the alternative of incorporating trade-offs in the area selection process (Stage 6) is preferred, and all relevant criteria have been incorporated into the trade-off analysis, then this stage is redundant.

The tenth stage is implementation of the conservation plan. This requires decisions on the most appropriate form of legal protection for each selected area and the most appropriate management actions for each selected area. An important consideration here is the scheduling of implementation. Resources are not normally available to act on all selected areas simultaneously. More vulnerable areas might receive priority, especially if the features they contain are absent or scarce in other areas (Pressey & Taffs 2001; Wilson *et al.* 2005). If it proves impossible to implement the plan because, for example, some areas are seriously degraded, budgets have changed or the forgone opportunity costs associated with parts of the plan are unacceptable to society, then it is necessary to return to Stage 6 and try again. Because of this inevitable scheduling problem we must accept that planning is a dynamic iterative process. Planners and policy-makers should return to earlier stages repeatedly because social and economic conditions change, social and political attitudes change, and knowledge accumulates. The plan that was right given the knowledge base and the social, economic and political climate last year will not necessarily be the best plan this year.

The eleventh and final stage is to monitor the effectiveness of management actions in sustaining the features that areas were selected for. Monitoring also requires that thresholds are defined, which, if passed, warn that unacceptable changes might be underway. The status of biological entities changes over time, as do social and economic conditions. Management actions that seemed appropriate at one point in time might be less effective at another point in time. Changes to management prescriptions are one response. As suggested above, another response might be to repeat the entire conservation-planning process periodically. The most desirable situation is that conservation planners have the facility to repeat the process as and when needed in order to take account of societal change and the gaining of new knowledge. Such iterative, dynamic managerial response with feedback is called "adaptive management" (Holling 1978).

1.4 Summary

The goal of systematic conservation planning is to identify areas that should have priority for the allocation of scarce biodiversity-management resources and to separate those areas from factors that threaten their persistence. Biodiversity is a complex concept that is impossible to estimate or quantify completely. For planning purposes we have to use partial measures or biodiversity surrogates that can be quantified and assessed in the field, for example taxa subsets, habitat types, etc. A key concept in systematic conservation planning is complementarity; the marginal contribution an area makes to representing the full set of attributes chosen as biodiversity surrogates.

Systematic conservation planning is a structured approach with feedback and reiteration, where needed, at any stage. In this book we recognize 11 distinct stages. They range from the engagement of people who influence, use and manage biodiversity in the planning region, through choices about how biodiversity is to be measured and mapped, to setting biodiversity goals that can be translated into quantifiable targets, estimating the extent to which targets have been met in existing conservation areas, using complementarity to identify new priority conservation areas, implementing planning choices on the ground and monitoring management actions to ensure the continued contribution by chosen areas to the conservation goal.

2 · Biodiversity surrogates

As we noted in Chapter 1 (Section 1.1), it is a truism that general bio-diversity is impossible to define, let alone quantify and measure in the field – the concept "biodiversity" is far too vague (Takacs 1996; Sarkar 2002; Sarkar & Margules 2002). In the abstract, the concept of bio-diversity includes diversity at every level of structural, taxonomic and functional organization of biota – as such the concept embraces virtu-ally all of biology. Traditionally, it has often been suggested that diversity should primarily be construed as diversity at three levels of the biolog-ical hierarchy: genes (alleles), species and ecosystems (Meffe & Carroll 1994). As a simplifying proposal in the face of intractable complexity, this convention has merit, especially if our concern is with conservation. If we conserve allelic diversity completely, we take care of much of the diversity within species. If we conserve all species, we conserve all entities at higher levels of the taxonomic hierarchy (though we may miss inter-specific hybrids). If we conserve all ecosystems we will conserve many communities and species. Nevertheless, even this catholic proposal does not capture all of the features of biodiversity (Sarkar 2002; Sarkar & Mar-gules 2002). In particular, it does not capture the diversity of ecological processes such as migrations. We will return to this issue of processes as components of biodiversity later (Section 2.7).

Acquiring complete inventories of genes, species and ecosystems, though it is now possible to map estimates of the last of these, is not a practical option in the short term. Land use change is proceeding apace and decisions have to be made now. Therefore, some surrogate or par-tial measure of biodiversity is required and it has to be mapped. This surrogate will be an attribute or some combination of attributes used to obtain information about biological diversity (however defined) in lieu of measuring biological diversity directly (Sullivan & Chesson 1993). These surrogates must satisfy two properties (Williams *et al.* 1994; Sarkar & Margules 2002): (1) *quantifiability* – they must be capable of quantitative assessment for use in conservation planning, that is, we must be able to say exactly how much of the surrogates have been adequately represented in a conservation plan; and (2) *estimability* – they must be realistically obtainable, for instance, from limited field surveys, remote-sensed data, models or combinations of these.

2.1 True and estimator surrogates

We need surrogates to represent biodiversity in planning protocols because the standard components of biodiversity (all genes, all species or all ecosystems) cannot usually be surveyed adequately within the temporal and budgetary constraints of a planning process (Austin & Margules 1986; Reyers *et al.* 2000). The term "surrogate" refers to an individual entity, such as a species or habitat type. Typically a planning process requires the use of a set of such surrogates, for instance, all species belonging to some taxon or all the habitat types in the planning region. Surrogacy is a relation between an "indicator" parameter and an "objective" parameter (sometimes called a "target" parameter; what we ultimately hope to conserve). An indicator parameter represents or replaces the objective parameter in planning protocols (Sullivan & Chesson 1993). In theory the objective parameter for biodiversity conservation should be all of biodiversity. Unfortunately, as noted earlier, that concept is impossible to define properly, let alone quantify and estimate. In practice this parameter is usually reduced to subsets of species or other taxa, for instance, species at risk (Sarakinos *et al.* 2001; Conservation International 2004).

Following Sarkar and Margules (2002), we call this objective parameter the "true surrogate" because it is intended to represent general (or the true) biodiversity in conservation planning. The important point to note is that any such choice of true surrogate must partly be made on conventional grounds (which does not mean that it is arbitrary). Consider three common – and plausible – candidates:

(1) *Character* or *trait diversity* (Vane-Wright *et al.* 1991; Faith 1992; Williams *et al.* 1994): The intuition behind this candidate is that evolutionary mechanisms typically impinge directly on traits of individuals in populations. The trouble is that "trait" is not a technical term within biology (Sarkar 1998). What we choose to call traits of organisms (or other entities) is largely determined by pragmatic considerations, for instance, what can be easily studied in a given research context. Trait diversity is, therefore, not precise enough to solve the quantification problem adequately;

(2) *Species diversity*: This choice can be made sufficiently precise and, in practice, it is the true surrogate that is usually invoked, often implicitly. Moreover, species diversity is the measure most commonly used in almost all practical discussions of biodiversity. This may well be because it is regarded as an adequate estimator surrogate (see below) rather than as a true surrogate. Nevertheless, it is important to note that species fall into the most well-defined category above the genotype in the taxonomic hierarchy and, because of that, the intuition that species diversity is a good true surrogate has some theoretical merit. The trouble is that we know very well that there is much more to biodiversity than the variety of species;

(3) *Species assemblage*, or *landscape pattern*, or *life-zone diversity*: These terms are used in different parts of the world by different people to mean similar things, though the spatial scale may vary. They reflect the intuitions that: (a) what is important is the variety of biotic communities with their associated patterns of biotic interactions; and (b) that focusing on communities will *ipso facto* take care of species since communities are composed of species. The chief disadvantage is that any classification of communities involves arbitrary decisions regarding what classes to recognize and what boundaries to draw between them (Haila & Margules 1996). However, although they are arbitrary, for many areas of the world, fairly precise and widely accepted classifications exist. For example, the life-zone classifications of Holdridge (1967) for central America, Ohsawa (1987) for the Bhutan Himalayas, land systems and bioregions of Australia (Christian & Stewart 1968; Laut *et al.* 1977; Thackway & Cresswell 1995), and similar classifications of eco-regions in Canada (e.g., Bastedo 1986) and world-wide (Olson *et al.* 2001), are often used. Because of this, for many regions, the quantification problem has a partially satisfactory solution. Nevertheless, only a small proportion of all species are used to define assemblages, bioregions or life zones and, therefore, they do not fully take into account species, or at broader scales even community, diversity.

However, in many planning situations, distributional data on the entire set of true surrogates are not available and cannot be collected given time and cost constraints. This problem is mitigated by the use of indicator parameters, hereafter called "estimator surrogates," which are intended to represent true surrogates in planning protocols. Given a set of true surrogates, whether a set of estimator surrogates adequately represents it is an empirical question (Landres *et al.* 1988). The use of estimator surrogates rests on an implicit assumption that there is a biological model linking the estimator-surrogate set and the components of biodiversity that form the true-surrogate set. For instance, if soil and climatic parameters are being used as estimator surrogates for plant species (the true surrogates), the implicit model claims that soil and climatic parameters determine plant species distributions. Before an estimator-surrogate set can be reliably used for conservation planning, the correctness of this model must be tested.

2.2 Establishing the adequacy of an estimator-surrogate set

Ideally, we should be able to predict the distribution of each true surrogate correctly from the distribution of the estimator surrogates. However, for any large true-surrogate set, to expect this amount of predictive success is unrealistic because: (1) very little distributional information may be available for many true surrogates; and (2) predictive niche models (see Chapter 3) are yet to achieve the required level of accuracy. Fortunately, when our purpose is limited to selecting conservation areas, a much more modest goal is sufficient. When we select conservation areas using estimator surrogates, we must ensure that the true-surrogate set is adequately represented. We review below the most common methods that have been used for this purpose. Other methods of estimating correlations such as the Mantel test have also occasionally been used (Ferrier 2002).

2.2.1 Species accumulation curves and indices

Ferrier and Watson (1997) suggested the use of species accumulation curves to study the performance of estimator surrogate sets (see, also, Ferrier 2002). As areas are added in an expanded network of potential conservation areas using the estimator-surrogate set, the cumulative representation of the true surrogates, such that each true surrogate is represented at least once, is recorded and graphed (see Figure 2.1). The representation achieved using the estimator surrogates can be compared to what would be obtained if true surrogates were used (the "optimum reference curve") and what would be obtained if sites are selected at random ("the mean random reference curve"). Ferrier and Watson defined a "species accumulation index" as $(s - r)/(o - r)$ where s is the log area under the estimator surrogate curve, r is the log area under the mean random reference curve and o is the log area under the optimum reference curve. The closer the index is to one, the better the performance of the

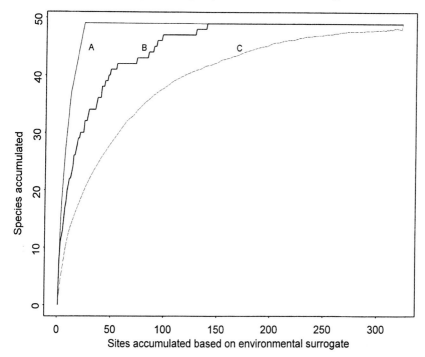

Figure 2.1. Species accumulation curve for New South Wales. In this case species are the true surrogates and environmental features are the estimator surrogates. A: the "optimum reference curve" that would be obtained if the planning units were selected using the true surrogates; B: the estimator surrogate curve; C: the "mean random reference curve," which results from the random selection of planning units. The data were from New South Wales; 429 invertebrate, 280 vertebrate and 2828 plant species were used as true surrogates (Plates 5 and 6). (Redrawn from Ferrier and Watson (1997). Reproduced with permission from the Department of Environment & Heritage, Australian Government.)

estimator-surrogate set. If the index is negative, the estimator-surrogate set is unacceptable because selecting sites at random would result in a better representation of the true surrogates. Bootstrapping can be used to estimate confidence limits for the index, and to test the statistical significance of differences in performance between different estimator surrogates (Ferrier 2002).

2.2.2 Surrogacy graphs

Surrogacy graphs, developed by Sarkar *et al.* (2000) and Garson *et al.* (2002a), are an extension and generalization of species accumulation curves. A species accumulation curve has the number of sites selected

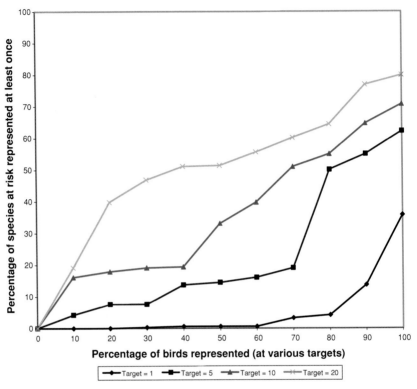

Figure 2.2. Surrogacy graphs for southern Québec. The estimator surrogates were 242 breeding bird species. The true surrogates were 400 animal and plant species at risk. For the true surrogates, the target was always one representation. The different graphs correspond to the different targets used for the estimator surrogates. (Redrawn from Garson *et al.* (2002a). Reproduced with permission from the Indian Academy of Sciences.)

as the independent variable and the number of species represented as the dependent variable. Surrogacy graphs are of two types. Both have the fraction of true surrogates that have met their target as the dependent variable. The salient innovation in the use of surrogacy graphs compared to species accumulation curves is that the target of representation for surrogates may vary, and need not simply be a single representation in a network of selected areas, as shown in Figure 2.2. In the first type of surrogacy graph, the independent variable is the fraction of estimator surrogates that have met their specified targets, and in the second it is the fraction of the total area so far selected. Each estimator and true surrogate may have a different specified target of representation depending on the biological and other requirements for its conservation. For instance, an endangered species may have a target close to 100% of all extant individuals whereas a common species may have a target of one population. The success of representation of true surrogates achieved in a surrogacy graph measures the performance of an estimator-surrogate set. If all true-surrogate targets are set to one representation, the second type of surrogacy graph reduces to a species accumulation curve. The same index and statistical tests used for species accumulation curves can also be used for surrogacy graphs.

2.2.3 Marginal representation contribution

Surrogacy graphs can be used only when surrogates have a specified target of representation. However, the use of targets in conservation planning (including area selection) has been criticized because most targets that have been used in practice do not have a firm biological basis (Soulé & Sanjayan 1998; see Chapter 6 [Section 6.3]). Consequently, it is of some interest to use a measure of the correlation between the contributions an individual area makes to representing true and estimator surrogates that is independent of target specification. The marginal representation index measures how much surrogate representation a cell adds relative to the total surrogate representation in all other cells (Sarkar *et al.* 2005). No target is required, but this is a measure of relative richness rather than complementarity – that is the price paid for avoiding targets.

Let Σ ($\sigma_j \in \Sigma$, $j = 1, 2, \ldots, n$) be a set of areas and Λ ($\lambda_i \in \Lambda$, $i = 1, 2, \ldots, m$) be a set of surrogates. For either type of surrogate, the marginal representation index, ρ_j, of area σ_j is defined by:

$$\rho_j = \frac{\displaystyle\sum_{k \in \Lambda} p_{kj}}{\left(\displaystyle\sum_{l \in \Sigma} \sum_{k \in \Lambda} p_{kl} - \sum_{k \in \Lambda} p_{kj}\right)} \tag{2.1}$$

provided that the denominator is not equal to zero, where p_{ij} is the expectation of finding λ_j (the j-th surrogate) at σ_i (the i-th area). (If the denominator is equal to zero, the data set contains only one area with data and every estimator surrogate is perfect.) The numerator consists of the expected number of surrogate occurrences in the site σ_j (Sarkar *et al.* 2005). The denominator is the expected number in all other sites. Thus, ρ_j provides a relative measure of the surrogate representation that σ_j adds to the other sites.

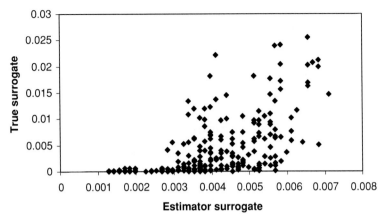

Figure 2.3. Marginal representation plot for Queensland. The landscape was divided into planning areas at a $0.10° \times 0.10°$ longitude × latitude scale. The estimator surrogate set consisted of 56 environmental classes. The true surrogate set consisted of 2348 plant species. The estimator surrogate set 1 described in Table 2.1 was used for this plot. The values along each axis were obtained using the formula given in Section 2.2.3. (From Sarkar *et al.* (2005). Reproduced with permission from Blackwell Publishing.)

Marginal-representation performance can be visualized as marginal-representation plots (Figure 2.3). These are two-dimensional plots of the true surrogate and estimator surrogate ρ_j values for each area in a region, used to analyze correlations between the marginal representations of the estimator and true surrogates. Visually, Figure 2.3 shows a non-random but non-linear association between the marginal representations of the estimator and true surrogates. The non-parametric Spearman's rank correlation test for non-linear correlations can be used to test for association. In Figure 2.3 there is a significant correlation ($p < 0.01$).

2.2.4 Spatial congruence analysis

The spatial congruence between cells selected based on the true-surrogate set and those selected based on the estimator-surrogate set is another measure of the performance of estimator-surrogate sets. Various coefficients of distance such as the Jaccard index or the Hamming distance between strings of zeros (absence) and ones (presence) can be used as quantitative measures of the congruence between the set of cells selected by true surrogates and that selected by estimator surrogates. However, most such indices ignore spatial correlations between the planning units. The statistical Syrjala test (Syrjala 1996), can be used to determine whether the spatial distribution of sites selected by true surrogates is identical to that selected by the estimator surrogates. However, this test is weak. For the data set described by Figure 2.3, Sarkar *et al.* (2005) found that the test did not reject the null hypothesis that the maps generated by the estimator- and true-surrogate sets were identical (significance level, $\alpha = 0.05$) even though it is clear from Figure 2.3 that the correlation is not perfect.

2.3 Traditional species-based surrogates

A variety of criteria have traditionally been used to select estimator surrogates for biodiversity in conservation planning, though usually without explicit consideration of the issues connected with the adequate representation of biodiversity that are discussed above. These surrogates have often been called "indicator" species in the literature (see, e.g., Caro & O'Doherty 1999). The criteria used to identify such surrogates include:

(1) Functional roles – some species, such as keystone species and focal species, are more important than others in ecological processes (e.g., plankton in the oceans and Spinifex in deserts). It has been suggested that the management of biodiversity should concentrate on such species (Walker 1992; Lambeck 1997, 1999; Fleishman *et al.* 2000), thus assuming that they are adequate surrogates for biodiversity. However, it is not clear that most ecological communities have keystone species (Power *et al.* 1996), although a possibility is that they do exist and often remain unrecognized because of the empirical difficulty of demonstrating their role in a community. Either circumstance forecloses the practical use of keystone species as estimator surrogates for conservation planning. Some studies have also suggested that keystone species, even if they can be identified, are poor estimator surrogates if threatened species are taken to be the true surrogates (Andelman & Fagan 2000; for more detail see [2] below).

The use of focal species for conservation planning goes beyond the use of keystone species (and umbrella species discussed below). It requires the identification of: (a) threatening processes, which endanger the focal species; and (b) a suite of several species rather than a single one. However, Lindenmayer *et al.* (2002) have pointed out that, in practice, it is difficult to identify the species most affected by each threatening process because of the lack of uniform data over all taxa (partly because of taxonomic bias – we generally have good data only on vascular plants, vertebrate animals and a few other taxa). Moreover, if being threatened is one of the criteria for identifying focal species, then we make an untested assumption of nestedness, that protecting the most threatened species inevitably protects others that are less threatened;

(2) Umbrella taxa – some species represent the "health" of an ecosystem (e.g., Chironomids) or the distribution patterns of other species, or require such extensive resources for their conservation that other species will be protected by default (e.g., top predators). Whatever intuitive appeal this idea may have, there has been little empirical evidence in support of the use of umbrella species in conservation planning. Andelman and Fagan (2000) explored the question of whether umbrella species are adequate estimator surrogates when the true surrogates are threatened species. As umbrella species, they considered big carnivores, habitat specialists, keystone species, species with a high age at first reproduction, long-lived species, those with the most available data, those that are most widespread and those that are most expensive to maintain. In a study at three spatial scales (that is, geographical ranges): (a) of the southern California sage-scrub community type; (b) of the Columbia Plateau that covers part of five north-western states in the United States; and (c) of the entire United States, they found that selecting areas using these estimator surrogates performed barely better than a randomly selected set of species used as estimator surrogates. In general keystone species performed worse than the others. However, using a variety of options for umbrella species, Fleishman *et al.* (2001) have reported more positive results;

(3) Iconic taxa – charismatic species (or other taxa) often perceived to be part of our cultural image (e.g., koala, giant panda, gorilla, etc.) are often chosen for immediate attention. Andelman & Fagan (2000) also studied the performance of charismatic species as estimator surrogates and found them to be no better than umbrella species;

(4) Conservation status – conservation planning should concentrate on species currently designated as endangered or threatened or with the potential to become at risk soon (Tracy & Brussard 1994; Sarakinos *et al.* 2001; Conservation International 2004). It is not controversial

that such species should be a special concern for biodiversity conservation. However, because they are features of biodiversity that are of obvious conservation concern, these species are better regarded as candidates for being true surrogates rather than estimator surrogates (Garson *et al.* 2002a) and because many of them are known and their locations can be mapped, when using threatened and endangered species as true surrogates it is not always necessary to invoke the use of estimator surrogates.

(5) Phylogenetic difference – species that are more phylogenetically distinct contribute more to total genetic and morphological diversity and so should be given priority for protection (e.g. Vane-Wright *et al.* 1991; Faith 1992; Weitzman 1993). An obvious way to operationalize this criterion for conservation planning is to use higher taxa instead of species for place prioritization. It is not controversial that diversity at levels other than the species level should be of interest for conservation planning. However, once again, this criterion is better used to choose true surrogates rather than estimator surrogates because it attempts to capture diversity at higher taxonomic levels than the species;

(6) Commercial importance – species with existing or potential commercial value (for harvesting, tourism, breeding new stock, pharmaceuticals, etc.) routinely have priority in conservation planning. This category may also include taxa which are sources of food and shelter for indigenous peoples. However, there is no reason to believe that these serve as adequate estimator surrogates since what is commercially important for human society may well have no correlation with what is important for the maintenance of biological variety – in this sense, such commercially important species are "arbitrary";

(7) Conspicuous species – the most conspicuous species (e.g., mammals, birds, butterflies, vascular plants) tend to be used because the data are available. This may be because they are large and obvious or it may be that good records exist because of keen amateurs, as in the case for birds and butterflies. Consequently, by myopic default, other species tend to be ignored. Again there is no reason to expect that these species will provide adequate estimator-surrogate sets – they may be "arbitrary" in the sense introduced above. In particular idiosyncratic (misfit) taxa, that is, taxa with special requirements (e.g., mobile wide-ranging species, colonial seabirds, vicariant species, etc.), may be overlooked (Ride 1975; Woinarski *et al.* 1992);

(8) Egalitarianism – all species should be included in conservation planning on an equal basis. Though this is often stated as a goal, it is currently impossible for the reasons given at the beginning of this chapter.

What deserves emphasis is that the performance of all these types of estimator surrogates should be explored using the techniques discussed above before they are recommended for use in conservation planning. Except for Andelman and Fagan (2000), who reported negative results for keystone and umbrella species, this has rarely been done. To the extent that the performance of taxonomic estimator-surrogate sets has been studied systematically, there is reason for scepticism about their adequacy.

2.4 Systematic surrogate sets

In recent years, conservation planning exercises have generally adopted a more systematic approach to selecting estimator surrogates. Surrogates are chosen with the expectation that they will sample the biological or environmental space uniformly, for example, all members of a taxon or set of taxa, the entire range of a set of environmental parameters, etc. We will restrict attention here to typical examples that have been widely used in conservation planning.

2.4.1 Sets of species and higher taxa

Most conservation planning exercises have routinely used sets of species covering entire taxa (all birds, all mammals, etc.) as estimator surrogates. These are a natural choice since it is reasonable to expect that uniform coverage over some taxonomically related class will cover other taxa because of ecological linkages. Most studies that have analyzed performance report pessimistic results about the adequacy of such taxonomic surrogate sets (e.g., Prendergast *et al.* 1993; Dobson *et al.* 1997; Flather *et al.* 1997; van Jaarsveld *et al.* 1998; Virolainen *et al.* 2000; Lund & Rahbek 2002). However, there have also been some optimistic reports. Moritz *et al.* (2001) present results showing that invertebrate taxa (insects and snails) perform well as surrogates for some plants and vertebrates in a tropical rain forest in Queensland, Australia. Example 2.1 presents positive results on the use of vascular plants as estimator surrogates. Panzer and Schwartz (1998) have also suggested that vascular plants might be an adequate estimator surrogate set if some invertebrates are used as the true surrogate set.

Example 2.1 Vascular plants as estimator surrogates: an example from New South Wales

Pharo *et al.* (2000) studied the use of vascular plant species as estimator surrogates for bryophyte and lichen species (the true surrogates) because the latter are difficult to identify at the species level. They studied 35 20 × 50 m² forested areas in New South Wales and analyzed two categories of vascular plants as estimator surrogates: all vascular plants (472 species) and overstorey species (dbh [diameter at breast height] >10 cm; 44 species). Bryophytes (78 species) and lichens (69 species) were sampled in the same sites using five quadrats (ranging from 20 cm² to 1 m²) on each of five different substrata (ground, logs, rocks, tree trunks and fallen branches). Each area had at least one unique vascular-plant species; consequently, representing all vascular-plant species at least once required the selection of all 35 areas. However, if it is only required that 90 % of all vascular plant species are represented, then 20 areas are needed. These 20 capture 51 bryophyte species (65 %) and 60 lichen species (87 %). Figure 2.4 shows the species accumulation curves. Eighteen areas are required to capture all overstorey species. These 18 captured 56 bryophyte species (72 %) and 61 lichen species (88 %). (Area selection was done using an algorithm based on complementarity which will be discussed in Chapter 5.) These are among the most promising results obtained for taxonomic surrogates.

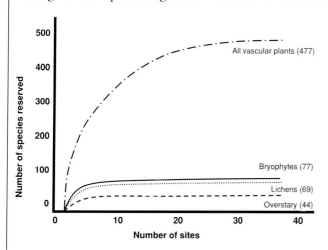

Figure 2.4. Species accumulation curves for plant, bryophyte and lichen data from New South Wales. For more discussion see the text. (Redrawn from Pharo *et al.* (2000). Reproduced with permission from Elsevier.)

2.4.2 Species assemblages

The term "assemblages" is used here generically to cover a range of classifications such as community, association, habitat type, etc. Assemblages are generalized biological entities more heterogeneous than taxa (i.e., they are not monophyletic). Taxa are distributed patchily within them and may only be present at particular times. Assemblages can be subjectively derived using a small number of dominant or prominent species or they can be generated from field data with numerical pattern analysis techniques to derive groups of co-occurring species (see Chapter 4).

Assemblages have the additional property of representing various alternative combinations of species and the interactions between them and, therefore, more ecological complexity than individual taxa. Larger organisms such as vascular plants and vertebrates, those most often used to delineate assemblages, interact with, and therefore encompass spatially, smaller organisms such as nematodes, arthropods, fungi, protozoa and bacteria, taxonomic groups which are usually ignored in conservation planning but are much more speciose than groups representing larger organisms (McKenzie et al. 1989). On the other hand, selecting an area or areas to represent an assemblage, which almost always consists of only a portion of that assemblage, is likely to miss some species and some ecological complexity, because it may be impossible to judge whether or not a given part of an assemblage is an adequate representation of the whole. Also, Araújo et al. (2004) found that the representation of vertebrates and plants achieved using assemblages was no better than that achieved by selecting areas at random in a data set from Europe, although their analyses were carried out at a very coarse spatial scale ($50 \times 50 \,\mathrm{km}^2$ grid cells).

2.4.3 Environmental classes

Environmental class is also a generic term covering land classifications based primarily on physical and climatic variables, which may or may not incorporate some biotic variables such as vegetation (Chapter 4). These classifications may be numerical or intuitive. Land Systems (Christian & Stewart 1968) are examples of intuitive classifications, while environmental domains (Mackey et al. 1989; Faith et al. 2001b) are examples of numerical classifications. Ward et al. (1998) show how similar classifications can be applied in aquatic environments. Different kinds of

Example 2.2   Species assemblages as surrogates in New South Wales

Oliver *et al.* (1998) tested the use of plant, vertebrate and invertebrate assemblages as surrogates for each other in north-eastern New South Wales. Data were recorded from 56 quadrats, 20 × 50 m² (0.1 ha) in size taken at the first point along transects previously used for wildlife surveys, bounded by latitudes 30° 00′ S to 31° 45′ S and longitudes 115° 15′ E to 152° 25′ E. Samples were taken from 32 unlogged and 24 logged areas. Ordination scatter plots were used to study the degree of similarity in assemblage composition among sites. Mantel tests were used to quantify and test the statistical significance of observed levels of similarity. Some of the analyses used subsets of taxa, distinguishing between understorey plants (dbh [diameter at breast height] < 0.1 m), overstorey plants (dbh > 0.1 m), birds, mammals and herpetofauna. Results were similar for both logged and unlogged forests. Very high correlations were found between all plant data and understorey plants, and vertebrate and bird data, but in both cases the data sets are not independent. Correlations between all vascular plants, vertebrates and invertebrates were also significant ($p < 0.0001$). For the subsets of taxa, there was low correlation between assemblages of birds, mammals and herpetofauna showing that these assemblages cannot be used as adequate estimator surrogates for each other. Understorey and overstorey plants showed a higher correlation but still not one that was statistically significant. Unfortunately, none of the methods for determining surrogate adequacy discussed in Section 2.2 were used.

environments are assumed to support different sets of species (with some overlap) and have been used at broad scales as estimator surrogates (e.g., Mackey *et al.* 1989; Richards *et al.* 1990; Belbin 1993; Pressey *et al.* 1996; Faith *et al.* 2001b). The measured relationships (similarity/difference) between localities used to classify environments numerically can also be used without a classification to describe environmental heterogeneity, and the resultant pattern may also be used as an estimator surrogate (Faith & Walker 1996a). The theoretical support for the use of environmental variables as estimator surrogates will be further discussed in Chapter 4 (Section 4.1).

There has been some controversy over exactly how environmental classes or variables should be used as estimator surrogates. Faith and Walker (1996a) advocated the use of a specific measure of environmental diversity ("Environmental Diversity" or "ED") as an estimator surrogate. The ED value of an area is calculated in an environmental ordination space. This value is the decrease in the summed distance from all points in the ordination space to the nearest point already selected for conservation. Maximizing the ED value is supposed to maximize true surrogate diversity. Araújo et al. (2001), using all available species in Europe as true surrogates, though only at a very coarse spatial resolution of 50×50 km^2, found that ED did little better in selecting sites than a random selection. They used a measure of ED that was slightly different from the one proposed by Faith and Walker (1996a) (see Faith 2003 and Faith et al. 2004 for responses, and Araújo et al. 2003 for a rejoinder).

However, ED is one among a number of measures of environmental diversity. Mackey et al. (1989) and Kirkpatrick and Brown (1994) used a hierarchical partitioning of environmental parameter classes as estimator-surrogate sets. Ferrier and Watson (1997) and Sarkar et al. (2005) used simple partitions of environmental parameter classes as estimator-surrogate sets. Sarkar et al. (2005) studied surrogate performance with data sets from Québec and Queensland. They also studied the performance of several subsets of these sets in an attempt to find an optimal surrogate subset – see Example 2.3.

Example 2.3   Performance of environmental surrogates in Québec and Queensland

Sarkar et al. (2005) analyzed data from Québec and the Wet Tropics bioregion of Queensland (Plates 7 and 8) at seven spatial scales (resolutions) ranging from $0.01°$ to $0.10°$ of longitude and latitude. For Québec they used 719 Plant and animal species (mainly species at risk) as true surrogates; for the wet tropics of Queensland, they used 2348 plant species as true surrogates. Areas were selected to form a potential conservation area network using the ResNet software package (Garson et al. 2002b).

Table 2.1 describes the partition of environmental variation in classes that was used to generate four different estimator-surrogate sets. For

Table 2.1 *Environmental estimator-surrogate sets for Québec and Queensland. An "×" indicates that the corresponding environmental class was used in this set. See the text for the methods of subdivision of the class used. Key: SL – soil type (four classes for Québec, six classes for Queensland, corresponding to the world soil map of the Food and Agriculture Organization [FAO]); ME – mean annual temperature (ten equal-interval classes for both data sets); MA – maximum temperature in the hottest quarter (40 equal interval classes for both data sets); MI – minimum temperature in the coldest quarter (40 equal interval classes for both data sets); MP – mean precipitation (ten equal-interval classes for both data sets); E – elevation (ten equal interval classes for both data sets); A – aspect (nine equal interval classes); S – slope (five classes based on standard deviations for both data sets). For further detail, see Sarkar et al. (2005)*

Estimator-surrogate set	SL	ME	MA	MI	MP	E	A	S	Total number of estimator surrogates	
									Québec	Queensland
1	×	×	×	×	×	×	×	×	56	54
2	×	×	×	×	×	×	×		51	49
3	×	×	×	×	×	×			42	40
4	×	×	×	×	×				32	30

this set they used soil type, slope, elevation, aspect and four climatic parameter types (annual mean temperature, minimum temperature during the coldest quarter, maximum temperature during the hottest quarter and annual precipitation). Annual mean temperature, precipitation, and elevation were divided into ten equal-interval classes. In general, finer divisions of environmental estimator-surrogate parameters should give better representation of true surrogates; however, the cost is increased computational complexity. The ten subdivisions were chosen because further refinement did not lead to significantly better results. The minimum temperature of the coldest period of the year and the maximum temperature of the warmest period of the year were divided into four equal-interval classes to ensure that each class had the same range as each annual mean temperature class. The use of equal intervals assumed that protecting sets of cells that contain all four classes will ensure that biotic features found in rare temperature regimes are adequately represented in a conservation plan. Slope

Table 2.2 *Landscape properties of the data for Québec and Queensland at the different spatial scales used in this analysis of surrogate performance. See the text for more details*[a]

	Québec		Queensland	
Scale	Number of cells	Average area km^2 (SD)	Number of cells	Average area km^2 (SD)
0.01°	33 967	0.850 (0.0190)	3 828	1.18 (0.005 19)
0.02°	23 474	3.38 (0.0788)	2 227	4.72 (0.0223)
0.04°	12 940	13.4 (0.322)	931	18.9 (0.0978)
0.05°	10 125	21.0 (0.506)	693	29.5 (0.158)
0.06°	8 156	30.1 (0.733)	518	42.4 (0.229)
0.08°	5 589	53.4 (1.31)	350	75.4 (0.446)
0.10°	3 890	83.3 (2.10)	251	118 (0.676)

[a] From Sarkar *et al.* (2005). (Reproduced with permission from Blackwell Publishing.)

was divided into five classes based on standard deviations. The use of standard deviations assumed that mid-range slopes are more important for biodiversity than extremes: in these regions this pattern was found for the richness of true-surrogate distributions. The soil data were divided into four classes for Québec and two classes for Queensland corresponding to the Soil Association types in the world soil map (FAO 1993). Aspect data were divided into the standard nine classes of 40° each. Preliminary results indicated that a finer division did not affect results.

The seven spatial resolutions used were: 0.01° of longitude × 0.01° of latitude, 0.02° × 0.02°, 0.04° × 0.04°, 0.05° × 0.05°, 0.06° × 0.06°, 0.08° × 0.08° and 0.10° × 0.10°. The number of cells with data decreased at coarser scales for both regions (Table 2.2). At coarser scales, cells were assumed to contain each of the surrogates represented at finer scales. Representation targets were set at 10% for both true and estimator surrogates.

Figure 2.5 consists of surrogacy graphs that describe the performance of environmental surrogate sets in Québec. The independent variable is the percentage of the estimator surrogates that achieve their targets. The results show a high level of success for all estimator-surrogate sets, which achieved above 90% representation of the true surrogates. However, random selection did equally well at this high resolution (0.01°). Figure 2.5b consists of surrogacy graphs that describe the performance in Queensland. In this case, the independent variable is the percentage of the area that has been selected. Note that at this scale (0.10°) in Queensland, the estimator-surrogate sets outperform the randomly selected cells.

Figure 2.6a,b show the spatial-scale dependence of the maximum representation of true surrogates for Québec and Queensland respectively. The marginal representation plot for Queensland at the 0.10° scale is depicted in Figure 2.3. For both Québec and Queensland, the Syrjala test did not reject the null hypothesis that the maps generated by the estimator- and true-surrogate sets were identical (significance level, $\alpha = 0.05$) even though Figure 2.6 shows an imperfect correlation.

Together, these results show that, at least for Québec and the wet tropics of Queensland, the use of environmental surrogates is a significant improvement over random selection of conservation areas at larger spatial scales (more specifically, at and above the 0.02° scale). Moreover, the routine achievement of a representation level of more than 90% of the true surrogate should be regarded as more than adequate when such coarse-grained environmental surrogate sets (only 56 classes for Québec and 54 for Queensland) are all that can be used for conservation planning. While the results show that estimator-surrogate set performance improves as spatial scale increases, the results are not as striking as those obtained by Garson et al. (2002a).

There is both theoretical and empirical support for using environmental classes as surrogates for at least some components of biological diversity. However, ecological and evolutionary history also play a role, sometimes a major role, in determining distribution patterns and this can only be traced by reference to particular taxa. A sensible strategy for prioritizing areas in this situation might be to aim to sample all environmental classes first, to include as many species as possible, and then to identify areas with species not adequately represented by that process for inclusion in priority areas.

(a)

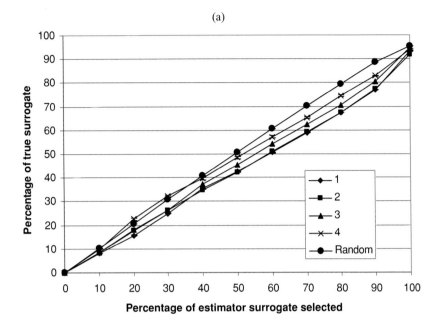

(b)

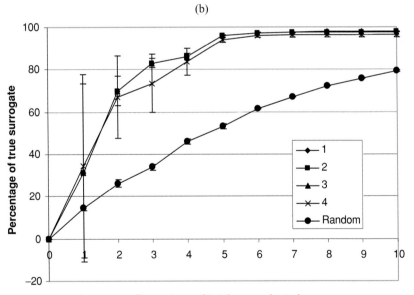

A network of conservation areas that represents the range of environmental combinations in a region is likely to encompass unknown species and known species with distribution patterns that are incompletely described. Furthermore, the data needed to delineate environments (e.g., climatic data, geological maps) are more generally available at a consistent level of detail across wide geographical areas, than are field records of species occurrences. On the other hand, as is the case with assemblages, protecting a single area as a representation of an environment is likely to miss some species because it is not clear what an adequate sample of an environmental class might be. Similarly, the relationships between environmental classes and the distribution and abundance patterns of taxa can be unclear and difficult to quantify, and some species may require a combination of environmental variables not recognized by a classification (Pressey 1992).

2.4.4 Combinations of surrogates

Taking into account the limitations on current knowledge, the limited resources for acquiring new data, the goal of adequately representing each true surrogate, and the assumptions that need to be made about the use of any of the kinds of estimator surrogates described above, it seems likely that in practice some combination of these estimator surrogates will be most widely applicable (see Chapter 8). In many localities some data on the distributions of taxa are available, but at an inconsistent level of detail and geographically biased. It may be that areas with endemic taxa, especially endemic vertebrates and plants, have been identified because these areas are inherently interesting to naturalists. Usually, at least some environmental data are available at a consistent level of detail and it may be

Figure 2.5. Performance of environmental surrogates. (a) Surrogacy graphs for Québec at the 0.01° × 0.01° longitude × latitude scale, using the percentage of the estimator surrogates selected up to the 10% target as the independent variable. (b) Surrogacy graphs for Queensland at the 0.10° × 0.10° scale, using the area selected as the independent variable. The "error" bars show standard deviations from 100 different runs rather than standard errors because the latter were too small to be depicted. These error bars underestimate potential uncertainty because only the uncertainty that arises due to the order of area selection is addressed. Uncertainty due to data quality is not addressed at all. The labels "1", "2", "3" and "4" correspond to the estimator surrogate sets described in Table 2.1. "Random" corresponds to the random selection of areas. (From Sarkar et al. (2005). Reproduced with permission from Blackwell Publishing.)

(a)

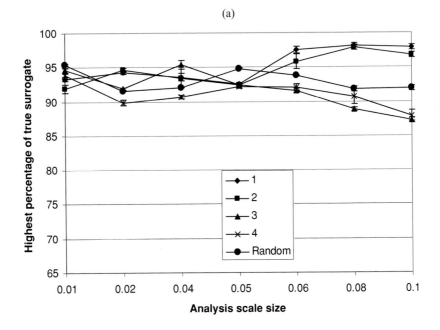

(b)

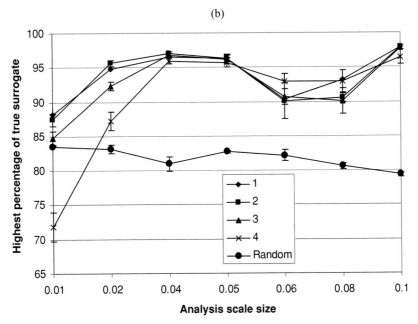

that assemblages have been mapped as, for example, vegetation or habitat types. One way to proceed in such cases would be to represent each environment or assemblage, overlay available distribution maps of taxa and/or areas of endemicity to see which, if any, were still not represented, and add areas to complete the representation of all possible surrogates (Nicholls & Margules 1993; Noss *et al.* 1999). Alternatively, the geographical locations of selected taxa, such as rare or vulnerable species, or areas of endemicity, could be used as seed points around which to build up a representation of each environment (e.g., Bull *et al.* 1993). There are problems with both of these approaches, though, because of the geographical bias of most field records, and because strict reservation may not be the most appropriate protection for rare or vulnerable species. Management to alleviate threatening processes, perhaps without changing the land tenure of the areas concerned, may be more appropriate for localized species of concern. Spatial modeling based on species localities can help alleviate the first problem, but it seems likely that some level of spatial bias will have to be accepted in most conservation planning exercises.

A method known as generalized dissimilarity modeling (GDM) (Ferrier 2002; Ferrier *et al.* 2002) is another more recent approach to combining both continuous environmental data and sporadic biological survey data. GDM is a non-linear matrix regression technique that statistically relates biological dissimilarities (or distances), measured between pairs of areas with records of species, to environmental differences (or distances) measured in terms of environmental variables with continuous spatial coverage. GDM addresses two common types of non-linearity encountered in ecological data. The first type is related to the well-established fact that the relationship between ecological separation of areas and observed biological (or compositional) dissimilarity is not linear, but rather is curvilinear. The Bray–Curtis dissimilarity metric used

---

Figure 2.6. Spatial scale dependence in (a) Québec and (b) Queensland. The *x*-axis is cell boundary size in longitude/latitude; *y*-axis is representation level of true surrogates when 100% representation of estimator surrogates has been achieved. The "error" bars show standard deviations from 100 different runs rather than standard errors because the latter were too small to be depicted. These error bars underestimate potential uncertainty because only the uncertainty that arises due to the order of area selection is addressed. Uncertainty due to data quality is not addressed at all. The labels "1", "2", "3" and "4" correspond to the estimator surrogate sets described in Table 2.1. "Random" corresponds to the random selection of areas. (From Sarkar *et al.* (2005). Reproduced with permission from Blackwell Publishing.)

most commonly in GDM is constrained between zero and one. As ecological and environmental distance increases, and areas share progressively fewer species, the observed dissimilarity will approach, but cannot exceed, a value of one. The relationship between ecological separation and compositional dissimilarity is therefore asymptotic. GDM fits models using generalized linear modeling rather than ordinary linear regression. This allows the curvilinear relationship between ecological separation and compositional dissimilarity to be accommodated by specifying appropriate link and variance functions. The second type of non-linearity relates to the rate of compositional change, or "turnover," along environmental gradients. With GDM, variation in the rate of compositional turnover along gradients is accommodated through automated non-linear transformation of environmental variables.

GDM predicts continuous patterns of compositional turnover across whole regions. These predictions can be used directly in conservation planning (e.g., Ferrier *et al.* 2004) or they can be used to derive classes using the numerical pattern-analysis techniques described in Chapter 4, which in turn can be used as biodiversity surrogates in conservation planning. Predictions of compositional turnover generated by GDM are also being used increasingly to assist in designing new biological surveys, by identifying locations that are likely to be most different from existing survey areas, in terms of biological and environmental composition (Ferrier 2002, Funk *et al.* 2005; and see Chapter 3, Section 3.3, Example 3.4, where survey gaps are identified using environmental data).

2.5 Surrogacy and spatial scale

The performance of estimator surrogates obviously depends on the spatial scale. At one extreme, if the entire region is considered to be one area – the lowest possible resolution – that area will contain all true surrogates and all estimator surrogates. Every estimator surrogate will have performed perfectly. At the other extreme, if areas are very small – the finest resolution – no area will have many of either the true or estimator surrogates. In this case, estimator surrogates will perform extremely poorly and there is no point in trying to use them. The interesting questions are whether there is a natural scale for any estimator-surrogate set and whether any such result generalizes over geographic regions. Spatial scale analysis, in the sense of a comparative study of the performance of estimator-surrogate sets at different spatial resolutions, remains relatively new in surrogacy analysis. Ferrier and Watson (1997) analyzed two Australian data sets at two spatial resolutions of 0.04 and 25.0 km^2. Andelman and Fagan (2000) considered data at different spatial resolutions, but did not perform any comparative analyses; their data at different resolutions came from different regions. Tognelli (2005) used all mammal species of Latin America as true surrogates, and several estimator-surrogate sets, including IUCN-listed (vulnerable, endangered or critically endangered) species, geographically rare species, flagship species and large mammal species. He analyzed data at the continental scale for Latin America and the national scale for Brazil, but the spatial resolution of the data (land unit size) remained the same.

Garson et al. (2002a) reported that the performance of birds used as estimator surrogates for species at risk improved up to an area size of about 100 km^2 but did not get appreciably better at coarser resolutions. This suggests that areas up to about 100 km^2 are the best to use for conservation planning with those particular estimator surrogates, but nothing more. Sarkar et al. (2005) analyzed the Québec and Queensland data sets at seven spatial resolutions as noted in the discussion of Example 2.3. They did not find any clear pattern of performance change with resolution. Much remains to be studied on this issue.

2.6   A protocol for the identification of an adequate surrogate set

If data for the selected true-surrogate set are available at a consistent resolution across the entire planning region, then there is no need for an estimator-surrogate set. However, if estimator surrogates have to be used (as is commonly the case) then, given the discussion above, we suggest the following protocol for their use and testing (Sarkar et al. 2005):

(1) Select a true-surrogate set and a group of candidate estimator-surrogate sets;
(2) Divide the planning region into cells of the appropriate size and project the region into an environmental space;
(3) Randomly select a set of locations (the calibration set) from the environmental ordination space, the larger this set, the better;
(4) Survey the corresponding cells in (geographical) space for the true- and all the estimator-surrogate sets;
(5) Construct surrogacy graphs for the sampled cells to determine the best or "optimal" estimator-surrogate set;
(6) Use the optimal estimator-surrogate set for conservation planning for the entire region.

In principle, this protocol can be carried out for any potential estimator-surrogate set without prior knowledge of its adequacy. However, such prior knowledge, for instance, knowing that a particular type of environmental-surrogate set is likely to be adequate, will help determine what group of candidate estimator-surrogate sets should be analyzed. In practical planning contexts, in the presence of temporal and budgetary constraints, this type of prior knowledge is critical for rational planning. Caro and O'Doherty (1999) have also argued that pilot studies always be carried out before any species sets are adopted as estimator-surrogate sets.

2.7 Diversity of ecological processes

So far, we have discussed surrogates for biological patterns, but there are also processes that form part of the concept of biodiversity. Many ecological processes hold the key to sustaining biodiversity into the future, for example, the movement of nutrients, energy, water and propagules throughout the landscape, both in space and time. We do not deal with this issue in this book, although one example described in Chapter 8 (Section 8.4) makes an attempt to take processes into account by using previous studies and local ecological knowledge to identify the sizes of conservation areas that would be needed to ensure the persistence of the processes, which included pollination, fire, plant–herbivore interactions, faunal migration and evolution (Pressey *et al.* 2003).

Some processes are clearly discernible spatially and these can be adopted readily for planning purposes when the data are available. Consider, for instance, the monarch butterfly, *Danaus plexippus*, which has two migratory populations in North America (Brower & Malcolm 1991). Beginning in late August, the eastern population migrates to Mexico for five months. These butterflies aggregate in millions in the high-altitude fir forests in the Sierra Transvolcanica, some 80 km west of Mexico City. There are about ten such overwintering sites all within an area of 800 km² on isolated mountain ranges between 2900 and 3400 m. Throughout the winter the butterflies remain sexually inactive. Survivors migrate north, starting in late March, and lay eggs on milkweed (*Asclepias* sp.) along the Gulf Coast of the United States. These eight-month-old remigrants die, but their offspring continue migrating north towards Canada. Two or three more generations are produced over the summer. By the end of summer, the last summer generation enters reproductive diapause and instinctively begins a southerly migration towards Mexico. The western population shows similar behavior, migrating to about 40 known over-wintering sites in California. What is striking about this behavior in both populations is that the migratory instinct is hereditary and yet so specific.

In California some measures have been taken to protect overwinter-ing sites but the sheer cost of land may result in only a very few of these sites getting the necessary protection. The future of the Mexican sites seems even bleaker. Until recently, the high altitude fir forests of Mex-ico had largely been spared from adverse anthropogenic effects. They now face at least six threats (Brower & Malcolm 1991): (1) large-scale legal and illegal logging for timber and firewood; (2) village expansion up the mountains; (3) increased use of fire to clear land; (4) invasion of the forests by lepidopteran pests; (5) spraying of *Bacillus thuringiensis*, an organic pesticide, the effect of which on monarch butterflies is unknown; and (6) increased tourism. It is quite likely that the overwintering sites in both California and Mexico will disappear and, with them, the monarch butterfly migrations. However, the disappearance of overwintering sites will not mean the extinction of monarch butterflies: there are numerous non-migratory tropical populations. What will disappear is the remark-able migratory behavior of the two populations discussed above which are an example of "endangered biological phenomena" (Meffe & Carroll 1994).

Biodiversity surrogates based on distributional patterns will not result in the conservation of such processes. Rather, the use of such surrogates must be supplemented with available knowledge of process patterns to identify many such conservation areas that are critical to their persistence.

2.8 Summary

General biodiversity is impossible to define precisely. Consequently, sys-tematic planning for biodiversity conservation must use partial measures or biodiversity surrogate sets. "True-surrogate" sets represent general bio-diversity in planning protocols and, because the latter term cannot be defined precisely, these must be chosen partly by convention and appeal to biological intuitions. In practice, "true" surrogates are usually subsets of species or other taxa such as endangered species or endemic species. How-ever, the distribution of even these true surrogates often cannot be fully assessed within the financial and temporal limits of a planning process. "Estimator-surrogate" sets can be used to represent true-surrogate sets in such situations. Whether an estimator-surrogate set adequately represents a true-surrogate set is an empirical question. Systematic methods exist to answer this question and should be used before an estimator-surrogate set is used to select a conservation area network.

Traditional surrogates such as focal, keystone or umbrella species have not performed well as estimator surrogates in most contexts where their adequacy has been examined. Systematic taxonomic surrogates such as all members of some taxa, for instance birds, mammals, etc., have only occasionally survived tests of adequacy. In contrast, estimator-surrogate sets consisting of environmental variables such as climatic variables, habitat types, etc., show promise if used with care.

Taking into account all the limitations imposed by our knowledge of biodiversity and biodiversity surrogates, a sensible approach for conservation planning is to use combinations of surrogates. Taxa such as endangered or endemic species and environmental- or habitat-type surrogates could be used concurrently to maximize the likelihood of representing general biodiversity.

3 · Data collection

Systematic methods for conservation planning require two separate but interdependent activities: compiling good data on the distribution and abundance patterns of the biodiversity surrogates to be conserved and the development of appropriate procedures for using those data to identify conservation-area networks. Planning procedures are discussed in Chapters 5–8. Data collection and databases are discussed in this chapter; data treatment will be discussed in Chapter 4.

A database for conservation planning must contain, at a minimum, information on what species (or other surrogates of biological diversity, such as species assemblages, habitats or environments – see Chapter 2) there are and where they are. Assigning conservation value to an area or set of areas using the principle of complementarity (Chapters 1 (Section 1.3) and 5) is essentially a matter of comparison between them and valid comparisons cannot be made unless this basic information is available. Supplementary information on abundances and accurate predictions of the responses of species and habitats to environmental changes is desirable, but rarely obtainable with the limited funds and time normally available for planning studies.

3.1 Areas and features

It is convenient to think of databases as consisting of areas and features. For conservation planning, areas are geographical units defined in space. They may be large or small, regular or irregular. Grid cells, catchments, habitat remnants and tenure parcels are some of the different kinds of areas used in conservation planning. Features should be viewed as properties of areas. Features are the biodiversity surrogates discussed in Chapter 2. They may be species or the characters of species, or more heterogeneous entities such as species assemblages, ecological communities, habitat types or environmental classes, etc. Data sets should be consistent in the surrogates they contain and in the measurement of those surrogates across the localities, regions, biomes, etc. that they cover because conservation planning requires comparisons between areas, as noted earlier.

This need for spatial consistency means that creating databases for conservation planning will usually involve some form of raw-data analysis. Data analysis may include one or more of the following tasks: classification of environmental variables into, say, environmental domains or ordination to derive patterns in continuous environmental space; classification or ordination of field locations of species records to derive, for example, species assemblages; and the estimation of wider spatial distribution patterns of species or assemblages with statistical or empirical models or machine-learning methods relating field records of occurrence to predictor (normally environmental) variables (see Figure 4.1 in Chapter 4). Some suitable analytical methods are described in Chapter 4.

Information about areas and surrogates is most conveniently recorded and stored in an areas × surrogates matrix. Figure 3.1 shows three types of areas × surrogates matrix. The first contains surrogate data of the "presence-only" kind where species, for example, have been recorded in some areas, but there is no indication of abundance or extent, and the lack of a recorded presence in other areas does not mean it is not there. It means that it has not been recorded there; so we do not know if it is there or not. The second matrix contains "presence–absence" data on surrogates. In this case, the absences are real within the limits of sampling intensity. This means that the surrogates were looked for and recorded as present where they were found or as absent where they were not found. The third matrix contains estimates of abundance (e.g., population size) or extent (e.g., percentage of cover) of the surrogates that are present with a zero abundance indicating absence. Conservation planning can be undertaken with all three kinds of data sets with, successively, more confidence in the results. Some real conservation plans have used data sets that were compilations of all three kinds (e.g. Pressey 1998; Ferrier et al. 2000).

Unfortunately, almost all records of species occurrences in herbariums and museums are of the presence-only kind. Historically, most field records have been collected opportunistically and the species collected are those of interest to the collector (Margules & Austin 1994). Most collections of field records map road networks. Figure 3.2, a map of Koala (*Phascolarctos cinereus*) records from New South Wales, Australia, illustrates this point. This map was compiled from museum records and from the results of a survey, which enlisted volunteers to send in records of sightings. The figure also shows part of the road network. Most records

Surrogates   Areas

(A)

	A	B	C	D	E	F	G	H	I	J	K	L	M	N	O	P
1	1	1	1	1	1	1	1	1	1	1	1	1	1	1		
2			1	1	1	1	1	1	1	1	1	1	1			
3						1	1	1	1	1	1					
4							1	1	1	1	1					
5							1	1								
6							1									
7							1									
8								1	1	1	1					
9								1	1						1	
10								1	1						1	
11										1						
12										1						
13												1	1	1		
14															1	1
15															1	1

Surrogates   Areas

(B)

	A	B	C	D	E	F	G	H	I	J	K	L	M	N	O	P
1	1	1	1	1	1	1	1	1	1	1	1	1	1	1	0	0
2	0	0	1	1	1	1	1	1	1	1	1	1	1	0	0	0
3	0	0	0	0	0	1	1	1	1	1	1	0	0	0	0	0
4	0	0	0	0	0	0	1	1	1	1	1	0	0	0	0	0
5	0	0	0	0	0	0	1	1	0	0	0	0	0	0	0	0
6	0	0	0	0	0	0	1	0	0	0	0	0	0	0	0	0
7	0	0	0	0	0	0	1	0	0	0	0	0	0	0	0	0
8	0	0	0	0	0	0	0	1	1	1	1	0	0	0	0	0
9	0	0	0	0	0	0	0	1	1	0	0	0	0	0	1	0
10	0	0	0	0	0	0	0	1	1	0	0	0	0	0	1	0
11	0	0	0	0	0	0	0	0	0	1	0	0	0	0	0	0
12	0	0	0	0	0	0	0	0	0	1	0	0	0	0	0	0
13	0	0	0	0	0	0	0	0	0	0	0	1	1	1	0	0
14	0	0	0	0	0	0	0	0	0	0	0	0	0	0	1	1
15	0	0	0	0	0	0	0	0	0	0	0	0	0	0	1	1

Surrogates   Areas

(C)

	A	B	C	D	E	F	G	H	I	J	K	L	M	N	O	P
1	1	10	15	21	13	11	19	10	11	15	33	32	23	41	0	0
2	0	0	67	45	98	83	44	19	16	51	11	14	45	0	0	0
3	0	0	0	0	0	34	56	76	87	34	56	0	0	0	0	0
4	0	0	0	0	0	1	78	52	45	92	0	0	0	0	0	0
5	0	0	0	0	0	0	60	56	0	0	0	0	0	0	0	0
6	0	0	0	0	0	0	42	0	0	0	0	0	0	0	0	0
7	0	0	0	0	0	0	31	0	0	0	0	0	0	0	0	0
8	0	0	0	0	0	0	0	66	78	99	65	0	0	0	0	0
9	0	0	0	0	0	0	0	18	0	0	0	0	0	0	5	0
10	0	0	0	0	0	0	0	23	56	0	0	0	0	0	3	0
11	0	0	0	0	0	0	0	0	0	78	0	0	0	0	0	0
12	0	0	0	0	0	0	0	0	0	89	0	0	0	0	0	0
13	0	0	0	0	0	0	0	0	0	0	0	45	32	46	0	0
14	0	0	0	0	0	0	0	0	0	0	0	0	0	0	77	87
15	0	0	0	0	0	0	0	0	0	0	0	0	0	0	22	98

Figure 3.1. Three kinds of surrogates × areas matrices showing (A) presence-only data, (B) presence–absence data and (C) abundance data. (Redrawn from Margules *et al.* (2002). Reproduced with permission from the Indian Academy of Sciences.)

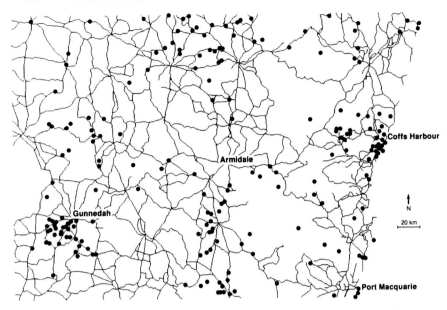

Figure 3.2. The distribution of Koala records in north-eastern NSW, Australia, in 1993. Most records mapped road networks or were clustered around towns. (From Margules & Austin 1994. Reproduced with permission from the Royal Society, UK.)

come from near roads or around townships. No systematic state-wide survey has ever been conducted of this very high profile charismatic species. There is no certainty that the parts of this map without records actually do not have koalas present, so the limits of its geographical range still cannot easily be defined. Thus, it is often not possible to define the limits of the distribution patterns of species because there are few, if any, records of where they were looked for and not found.

The reality is that existing data sets are generally far from ideal. In the face of numerous and probably irreversible planning and policy decisions affecting biodiversity being made every day, it is necessary to make full use of existing data. However, it is also necessary to acknowledge the limitations of these data and establish ideal data requirements, not only as an aspiration, but also because identifying the ideal helps to capitalize on the useful information content of existing data (Williams *et al.* 2002).

More systematic methods for the collection of new field data of the presence–absence kind are described in Section 3.3 below. These methods are still not widely used, possibly because they tend to be more labor-intensive, but more likely because most field records are collected for purposes other than conservation planning. For example, when data are collected for systematics, those doing the collecting are typically more interested in the organisms themselves than in where they come from. Meanwhile, the best use must be made of existing data, even if field records are geographically biased and incomplete. Some techniques for adding value to museum and herbarium collections by estimating distribution patterns from presence-only data are described in Chapter 4.

3.2   Sources of data

The task of collecting, determining systematic affinities, and naming all species via traditional systematics, is daunting. Scientific names have been given to approximately 1.4 million species of plants, animals and microorganisms (Wilson & Peters 1988; Erlich & Wilson 1991) but this is only a fraction of all species. Estimates of arthropod diversity alone range from 5 million to 30 million species (Erwin 1982, 1983; Stork 1988; Ødegaard 2000) and the other invertebrate phyla are even more poorly known. Estimates have been made that if the collection and description of new species was to continue at the current rate, using traditional methods, it would take several thousand years to catalog the world's biological diversity (e.g., Disney 1986; Soulé 1990) and in fact the rate is slowing because funding for taxonomy has declined (e.g., Stork & Gaston 1990; Whitehead 1990; Richardson & McKenzie 1992).

There are many organizations throughout the world holding data that can validly be used for conservation planning. Existing biological data can be extracted from collections in museums and herbariums, from various departments of government such as natural resource management agencies, and from non-government organisations.

Data have to be extracted from the institutions holding them before bias can be assessed and decisions taken either to proceed with existing data, collect new data, model spatial patterns (Chapter 4) or reject the data. This is usually a time-consuming first step in conservation area network planning. There is no way to avoid this step and there are no short cuts. Expensive time-consuming new surveys are the only and, when existing data are unsuitable, necessary alternative. Fortunately, digitized data sets are becoming widely available. Ready access to existing data sets is not always possible. Custodians may place restrictions on access to their data, or charge for the use of data. In some cases, access is restricted to protect the locations of rare or threatened species (e.g., Gibbons *et al.* 1993; Sarakinos *et al.* 2001 – see Chapter 8 [Section 8.2]). In other cases, custodians may wish to protect ownership for research purposes. Belbin *et al.* (1994), in a review of data holdings in Australia, recognized four categories of data access: unrestricted access, formal acknowledgment required, permission required (selective or incomplete access), and confidential. They found that 51 % either required acknowledgement, were confidential or required permission, and 49 % were unrestricted.

Environmental data are more widely available, more accessible and generally exist in a more consistent form than most biological data. Environmental data alone can be used as surrogates for biodiversity and they are required for any formal modeling of wider distribution patterns of species or populations from the point records that field collections represent (see Chapter 4). Environmental data fall into the three broad categories: terrain, climate and substrate. Terrain refers to surface morphology and includes parameters such as elevation, slope, relief and aspect. These parameters can be recorded from topographic maps if the resolution is sufficiently fine, but in most cases it will be more appropriate to interpolate them from digital elevation models (DEMs). A DEM at a resolution of 1 km^2 is now available for the entire globe (USG Survey 1998). Globally, DEMs at more detailed resolutions are not yet routinely available or as accurate as topographic maps. The construction of a DEM can be time-consuming and is technically demanding. However, DEMs allow consistent and repeatable interpolation across whole regions and should constitute the necessary first step in generating environmental surfaces (Hutchinson 1993). Climate data are available from national meteorological bureau(s), but may have to be digitized for spatial modeling. Climate data can be interpolated spatially with the aid of DEMs by fitting surfaces as smooth tri-variate functions of latitude, longitude and elevation (Hutchinson 1995). Climate data can sometimes be augmented with data collected by forestry, conservation, agriculture and water resource agencies. Physical and chemical substrate data may be the most difficult to obtain. However, substrate mapping has been completed in a number of countries, regions and biomes. For example, thematic maps of lithological substrate, soils and landforms may be available.

Maps are a popular and efficient way (measured both by the information/ink ratio and speed of communication) of summarizing and communicating existing data (Tufte 1990). Two classes of map relevant to computer aided (GIS) mapping are referred to as vector maps and raster maps (Burrough 1986). The former define areas by joining boundary points as polygons, often of irregular shape and size. The latter define a regular array or grid of areas of similar shape and size. Vector maps allow

Example 3.1 A database for conservation planning in Papua New Guinea

Papua New Guinea (PNG) occupies the eastern half of the large tropical island of New Guinea and its associated off-shore islands which lie directly north of Australia. Local relief has extremely high variation. Approximately 70% of the country is covered in tropical forest. The other 30% is a mosaic of village agriculture, grasslands, and secondary and primary forest (Plate 4). In 1996, the Global Environment Facility (GEF) sponsored a study to identify biodiversity priority areas in PNG, which was to serve as an in-country evaluation of the BioRap conservation planning tools developed earlier (Margules *et al.* 1995), as well as assist PNG with its biodiversity planning and management. The database used in this study is described here (see Nix *et al.* 2000; Faith *et al.* 2001b) and the data treatments to derive the biodiversity surrogates are described in Chapter 4.

The biodiversity surrogates were environmental domains, vegetation types and the small number of rare and threatened species on the PNG government's official list of rare and threatened species. Nix *et al.* (2000) and Faith *et al.* (2001b) both describe the derivation of ten species bioclimatic profile clusters. A bioclimatic envelope model was produced for each of 87 species using BIOCLIM (Chapter 4; Example 4.1). These models were then grouped into ten classes. The plan was to include these as biodiversity surrogates, but it was found that they were so widespread that they were easily represented in all possible combinations of conservation areas; consequently, they were not included in the priority area analysis described later in Chapter 8 (Section 8.5). It was also the intention to use more species to derive the plan described in Chapter 8. However, approximately one third of the funding and one third of the time for the project was spent on searching museum and herbarium records to find that only 87 species had five or more accurately geo-referenced locations. There are undoubtedly many more, but a decision was taken at that stage to cut the losses, not spend more valuable time searching collections and to proceed with environmental domains, vegetation types, and the rare and threatened species. The biodiversity wealth of PNG is currently estimated to be 5% of the world's total (Graham *et al.* 2005). Balanced planning cannot wait for a full and accurate assessment. The features listed above represent biodiversity surrogates that could be assessed within 12–18 months and therefore could be used in a rational allocation of scarce land resources between competing demands of agriculture, forestry and conservation (Faith *et al.* 2001b).

Resource mapping units and PNGRIS

The Papua New Guinea Resource Information System, known as PNGRIS, contains information on natural resources, land use and human population densities for the whole country (Bellamy & McAlpine 1995; Keig & Quigley 1995). The land units for which this information is recorded are called resource mapping units (RMUs). These land units are widely used by government agencies in PNG and were, therefore, adopted as the candidate conservation areas for this study. RMUs were mapped from aerial photographs during extensive land-resource surveys carried out by the Australian Commonwealth Scientific and Industrial Research Organisation (CSIRO) in the 1970s. They vary in size from 0.45 km^2 to 8508 km^2. All environmental and biological data used for this study were assembled and allocated to RMUs.

A forest inventory mapping (FIM) system has evolved from PNGRIS. It is a national coverage at a scale of 1:100 000 showing:

- Areas of forest by forest type and non-forest, including land use, as at 1975;
- Indicative merchantable species composition and gross stocking rate for each forest type;
- Areas logged between 1975 and 1996;
- Forest areas cleared and converted to other forms of land use between 1975 and 1996;
- Existing and proposed forest and timber concession areas;
- Gazetted protected areas;
- Forest areas excluded from logging due to extreme physical limitations such as slope and susceptibility to inundation.

This database was used to derive the vegetation types as described below. The vegetation types were then incorporated into RMU descriptions.

Environmental Data

The digital elevation model

The DEM was constructed from digital spot-height data, elevation contours, lakes, streamlines and coastlines from the 1:1 m scale Digital

Chart of the World supplemented by elevation data from the 1:1 m scale Australian Geographical Series and the 1:0.5 m scale Tactical Pilotage Charts, using the ANUDEM software package (Hutchinson 1989; 1997). It has a resolution of 0.01°, or approximately 1 km². The overall error was estimated to be 70 m, although in areas of complex topography, it could be as large as 200 m (Nix et al. 2000; Faith et al. 2001b).

Climate surfaces

Data from the Australian Bureau of Meteorology and the PNG National Weather Service, with additional rainfall data from McAlpine et al. (1975, 1983), were used to fit elevation-dependent thin-plate smoothing spline surfaces describing monthly mean daily maximum and minimum temperature and rainfall. Grids of monthly mean and annual mean temperature and precipitation were obtained by coupling these surfaces to the DEM.

Although there were only 101 temperature data points, the strong dependence of temperature on elevation meant that the temperature surfaces have errors of only 0.5 to 1 °C. There is a denser network of rainfall stations, giving 560 precipitation data points. The spatial complexity of precipitation requires a denser network. Errors in the monthly mean precipitation surfaces were estimated to be between 10 and 20 % (Nix et al. 2000).

Lithology

Lithology was digitized from the 1:250 000 geological maps of PNG. Over 900 lithology classes were present on the original maps. These classes were grouped into the 23 that are listed in PNGRIS by H. A. Nix and J. L. Stein.

Biodiversity surrogates

As described in Chapter 4, the climate, terrain and lithological data were used to derive environmental domains and the vegetation types were re-analysed to derive new vegetation types. Thus, the biodiversity surrogates used to identify priority areas in PNG were environmental domains, vegetation types and the PNG government's 1997 list of rare and threatened species.

data to be presented at any scale. However, vector maps merely give the impression of lacking scale: the true resolution is limited, as always, by the number and precision of data points. What is important is that data are registered with precision, either as "points" with small errors, or on fine grids, so that they can be used subsequently in analyses at a broad range of spatial scales depending on the questions to be addressed. In this regard, there is no fundamental difference between vector maps and raster maps.

3.3 Collecting new data with field surveys

Records of the localities of species, the raw survey data that form the basis for so much of biodiversity conservation planning, have to be collected in the field and the methods of collecting, the survey design and sampling strategy, have profound implications for the subsequent use of those data. New biological surveys, including future collecting expeditions that museums or herbariums might mount, can be greatly improved for the purposes of conservation area network design with the application of some systematic procedures and the recording of presences and absences.

Biological surveys are necessarily samples; samples of the biota and samples of the geographical regions in question. They are usually conducted over areas defined by tenure, biogeographical boundaries or boundaries associated with major ecological processes. Most are relatively large heterogeneous areas. Even local surveys for specific purposes such as environmental impact assessment must be concerned to some extent with the wider region in which they occur or there is no basis for comparison and, thus, no way of assessing relative importance or value.

For conservation planning, survey data must be accurate, fairly representing the true distribution patterns of the species recorded. Biological survey design is a neglected topic (Austin & Heyligers 1991). Surveys themselves can be tedious, time consuming and labor intensive. Biological survey is not often seen as a scientific endeavor and is therefore ignored by textbooks. Yet, as noted earlier, the design of a survey is important for future data analysis and interpretation: rigorous design rules should be explicated and applied.

3.3.1 Theoretical issues

Survey design needs a conceptual framework based on ecological theory (Margules & Austin 1994). Consider the problem of mapping the

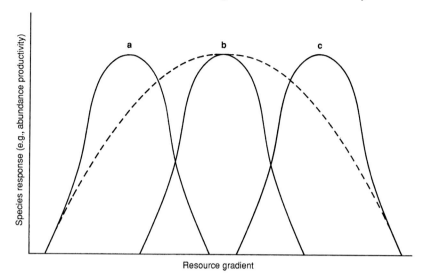

Figure 3.3. A theoretical model of the realized niches of three species, a, b & c, and the fundamental niche of b in the absence of competition from a and c. (From Austin & Smith (1989). Reproduced with permission from Kluwer Academic Publishers (Springer Science).)

distribution patterns of species. They are not random or uniform, but controlled or restricted in range by a set of variables interacting in complex ways. Each species has a unique distribution determined by its genetic characteristics and physiological requirements and by its ecological interactions, such as competition and predation, with other species. The resultant spatial pattern shows high population densities at scattered centres of distribution and broad areas of overlap at lower densities. Plant ecologists have adopted the continuum concept to explain this pattern. Animal ecologists use the concept of a niche.

There are a number of authoritative texts on plant community ecology and the niche concept (e.g., Whittaker 1972; Giller 1984), as well as the classical papers by Hutchinson (1958) on the niche and Bray and Curtis (1957) on a vegetation continuum. More recently, Austin (1985) and Austin and Smith (1989) have elaborated a theory of the ecological continuum with some testable hypotheses, and identified similarities with the concept of a niche.

Figure 3.3 illustrates the niche and continuum concepts. The figure shows the realized niches (the actual portion of a resource gradient occupied in the presence of competitors) of three species, and the fundamental

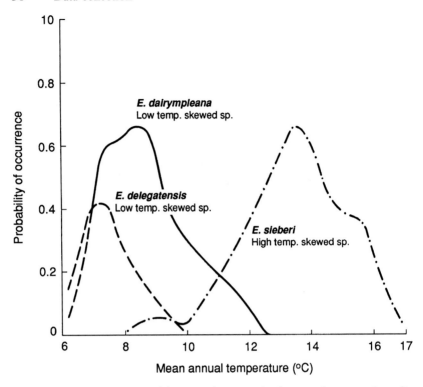

Figure 3.4. Response curves of three *Eucalyptus* species from south-eastern Australia to temperature. These curves are asymmetrical and show that species respond differently to resource gradients. (Courtesy of M. P. Austin.)

niche, that is, the portion of the resource gradient that can be occupied in the absence of competitors, of one species. The shapes of the realized niches, or response curves, are shown as symmetrical. Intuitively, it seems unlikely that they would be symmetrical in the real world, but there are few empirical studies of the shape of a realized niche. One exception is by Austin (1987), who showed that the shapes of the response curves of *Eucalyptus* tree species to temperature in south-eastern Australia were indeed asymmetrical (Figure 3.4).

The shapes of species responses become even more complex when plotted against two or more environmental variables. Figure 3.5 shows the complex response curve of *Eucalyptus pauciflora* in relation to rainfall and temperature. This species occurs near the tree line as well as fringing frost hollows in areas of cold air drainage in generally warmer regions, which is why it is found, though less frequently, at warmer annual temperatures.

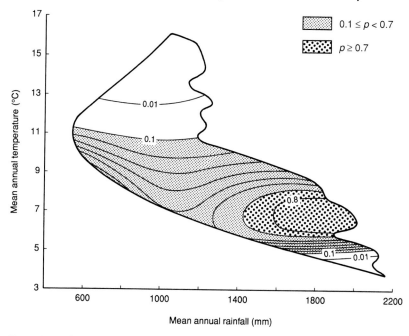

Figure 3.5. The modeled occurrence of *Eucalyptus pauciflora* in temperature and rainfall space. Rainfall/temperature combinations that fall outside the envelope illustrated do not occur in this region. This species occurs in south-eastern Australia near the tree line as shown in the bottom right-hand corner where it drops off rapidly as temperature decreases. It also occurs at higher mean annual temperatures, between about 7 and 8 °C, where it fringes lines of cold-air drainage which create frost hollows. These localized conditions are masked by mean annual temperature. In effect, this represents an inverted tree line. (From Austin *et al.* (1990). Reproduced with permission from the Ecological Society of America.)

The underlying causes of these patterns are still a matter of conjecture. While it is clear that temperature is limiting at one end of the gradient in Figure 3.4, is it also limiting at the other end, or is it the presence of competitors, or are the other species simply better adapted to warmer temperatures, or is it some combination of all three? *Eucalyptus pauciflora* is cold adapted, as clearly shown in Figure 3.5, but would it expand beyond the fringes of frost hollows in the absence of competition from other species at the warmer end of its distribution? While these questions are intriguing, for the present purpose the answer is not important. The fact that, on the whole, species respond differently – in the terminology of Gleason (1922), individualistically – to environmental gradients, provides

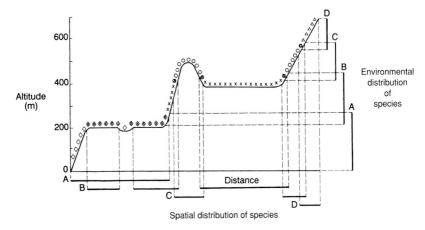

Figure 3.6. Patterns of co-occurrence of four species along an altitudinal gradient. A plant community is a landscape concept and the recognition of communities depends on the frequency of environmental combinations in particular landscapes. See text for a detailed discussion. (From Austin & Smith, 1989. Reproduced with permission from Kluwer Academic Publishers (Springer Science).)

the first and most important rule of survey design: stratify the region to be surveyed using major environmental variables.

3.3.2 A comment on the community versus continuum concepts

The continuum concept of vegetation organization is now a well-established doctrine of plant ecology, appearing regularly in textbooks (e.g. Mueller-Dombois & Ellenberg 1974; Krebs 2001). However, the notion that plants and animals occur in co-evolved communities persists. While there are numerous examples of pairs of co-evolved species – insects and host plants provide many of them – the idea that whole communities have co-evolved is more problematical. There is certainly a practical need to distinguish biological entities from one another at a higher order of generalization than the species, but there is no need to invoke co-evolution of whole assemblages to do this. If communities can be thought of as collections of co-occurring species rather than "superorganismic" entities of co-evolved species, then spatially distinct communities can be readily defined, as described in Chapter 4.

The argument for regarding communities as assemblages of co-occurring species was clearly explained by Austin and Smith (1989) using the diagram reproduced here as Figure 3.6; a transect up a mountain on which four species are present. The communities that might be recognized on this mountain, based on the frequency of species occurrences, are A, AB, B, C and D. The rare combinations BC and CD would probably be regarded as ecotones. These communities are no more than a consequence of the spatial pattern of the landscape. The altitudinal gradient reveals a continuum of species with A being replaced by B, B by C and C by D with increasing altitude. Different communities might be recognized on an adjacent transect with the same four species, but a different landscape pattern. If the first bench in Figure 3.6 was 30 m lower, AB would not occur because at 170 m only species A is present. Similarly, if the second bench was 30 m higher, BC would become a community rather than a rare ecotone. The communities on this new transect would be A, B, BC, C and D.

Recognizing and describing such communities is useful for planning, management and communication, but they cannot be extrapolated to other regions unless those regions have the same landscape pattern on the same rock type with the same climate, in which case they would probably be part of the same region.

3.3.3 Statistical issues

Sampling strategies for biological surveys have to be designed with the use of the data in mind. The required information consists of distributional patterns of species or communities and the way these patterns vary across a region rather than an accurate measure of central tendency such as a best estimate of the mean of, say, the abundance of a species or the number of species (Austin & Heyligers 1989).

Two appropriate analytical techniques described in Chapter 4 are multi-variate pattern analysis and spatial modeling. Pattern analysis (e.g., Anderberg 1973; Belbin 1991; Faith 1991) provides an explicit quantitative description of patterns in survey data. Spatial modeling allows the prediction, with an estimate of reliability, of survey records to unsampled places. Common and useful examples are heuristic models (e.g., Nix 1986; Walker & Cocks 1991; Carpenter et al. 1993), statistical models (e.g., Austin et al. 1984; Austin et al. 1990; Nicholls, 1989, 1991) and computational methods, which include maximum entropy techniques (Phillips et al. 2004) as well as neural networks (e.g., Hilbert & van den Muyzenberg 1999) and genetic algorithms (e.g., Stockwell & Noble 1992; Stockwell & Peters 1999).

The successful application of these techniques depends on an adequate environmental stratification to ensure that the range of variation across a region is sampled. This is important because then the prediction of records to unsampled sites and the description of patterns is an act of interpolation rather than extrapolation beyond the range of the data.

Underlying both pattern analysis and statistical modeling are the assumptions that samples are drawn from the same population, that the sample is unbiased within any stratum and that samples are comparable, that is, of the same size and recorded in the same way. Thus it is necessary to adopt statistical rigor. In practice, this means that samples should be recorded from a known plot size, preferably the same size, though small variations can be accounted for in an analysis, and, within the lowest level of stratification, from a random location.

3.3.4 Practical issues

Resources for surveys, funds, personnel and time are invariably limited, so these constraints have to be built into survey designs. It is necessary at this stage to apply common sense.

In practice, this means choosing variables for stratification that can be mapped and selecting field-sample sites that can be reached. Since the purpose of a survey is the description of ecological patterns and not necessarily the discovery of the processes that create those patterns, reasonable leeway exists in choosing variables for stratification. There are usually a number of geographic locations representing any one environmental stratum so it is sensible and efficient to choose the most accessible for field sampling. Occasionally, a disproportionate effort will be required to reach sites representing some rare environments, but the incidence of these can be minimized.

These practical issues should guide survey design after theoretical and statistical principles have been applied. They may, and often do, demand a relocation of some survey sites. Some of the examples below show how practical constraints can be incorporated explicitly into survey design without compromising the principles of environmental stratification or statistical rigor too greatly.

The examples below have been chosen deliberately to cover a wide range of situations; different scales, different environments and different resource bases to support the surveys. To some extent, every survey is different and will therefore require a unique approach, but there are commonalities embodied in the theoretical, statistical and practical issues raised above and the examples below have also been chosen deliberately to emphasize both differences and commonalities.

Example 3.2   Coastal hardwood forests in northern New South Wales, Australia

Austin and Heyligers (1989, 1991) conducted a survey of tree species in hardwood *Eucalyptus* forests on the north coast of New South Wales, Australia. Their objective was twofold. First, they wanted a floristic description of the forests that could provide a suitable database to inform land-allocation decisions and forest management and at the same time evaluate incremental improvements to vegetation survey design. Second, they wanted to provide a suitable stratification for a survey of the vertebrates of these forests. Braithwaite *et al.* (1983, 1984) had found previously that different forest tree assemblages supported different numbers and densities of arboreal marsupials and birds. Braithwaite and his colleagues wanted to test this finding in a new area (Braithwaite *et al.* 1989) and extend their coverage to ground-dwelling mammals.

The survey area was the catchments of the Bellinger, Nambucca, Macleay, Hastings and Manning Rivers (Fig. 3.7), a total area of approximately $20\,000\,\text{km}^2$. These rivers originate on tablelands and flow through deeply dissected mountains and hills onto a coastal plain and then into the Tasman Sea. Most of the area remains forested, though substantial clearing for agriculture and settlement has occurred in the broader valleys and on the coastal plain.

One design tool for surveying large heterogeneous regions efficiently, while incorporating the principle of environmental stratification, was proposed by Gillison and Brewer (1985). This is the gradient-directed transect or *gradsect*. The idea is to identify a set of transects which intercept the major environmental strata. If these transects are aligned along gradients of steep environmental change, then the greatest amount of environmental variation can be intercepted in the shortest distance. Austin and Heyligers proposed refinements to the idea, including replications, where possible within transects, and explicit rules for identifying actual field-sampling sites, which incorporate another lower level of environmental stratification.

Austin and Heyligers' design protocol consisted of seven steps:

(1) Identify the major environmental variables influencing the distribution patterns of vegetation;
(2) Recognize the subset of environmental variables best suited to determining the position and direction of gradsects;

LOCATION

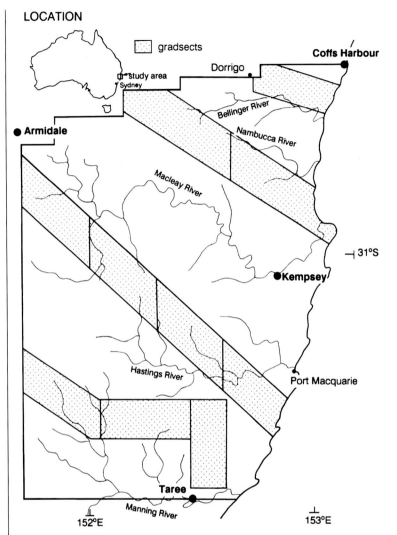

Figure 3.7. The location of the study area on the north coast of New South Wales and the locations of the four gradsects within the study area. (From Austin & Heyligers (1989). Reproduced with permission from Elsevier.)

(3) Choose the best available data for environmental stratification and the best available technology for implementing gradsect selection;
(4) Stratify the environment within gradsects and break the gradsects up into segments to allow replicate sampling of the same environmental strata at different geographical locations;

(5) Decide whether or not another level of stratification is needed to take account of environmental variation at the local scale;

(6) Decide on the degree of effort that should be allocated to sampling the rarest environmental strata as opposed to increasing replicate sampling of the common strata; and

(7) Be flexible. Some sample sites selected in the laboratory will be useless because, for example, they have been cleared of native vegetation or access is denied. New sites have to be chosen following established rules. Also, as happened in this survey, there may be a failure to identify, and therefore stratify on, an important environmental variable. Again, new sample sites have to be chosen following established rules.

Each of these steps requires subjective decisions to be taken. Such decisions have to be informed by experience, intuition, local knowledge and common sense. In that sense, "objective" survey design may be an oxymoron. However, as long as the procedure is explicit it is repeatable and can be improved incrementally with each new survey.

Survey design

Austin *et al.* (1984) had shown previously that rock type, rainfall and temperature have a strong influence on the distribution of tree species at this regional scale. As temperature correlated strongly with altitude, altitude was used because it could be easily determined from topographic maps.

Data for rainfall, altitude and rock type were recorded on a regular grid of points of $0.01°$ (approx. $1\,km^2$) spacing. Maps of the environmental cells formed by all possible combinations of classes of environmental factors were created and gradsects were selected, which together represented as many environmental cells as possible. Figure 3.7 shows the gradsects that were selected and Figure 3.8 helps illustrate this process. The stars represent rainfall and altitude combinations covered by three gradsects and the open circles represent new combinations covered by the addition of the fourth gradsect, which is the short northern one in Figure 3.7.

The dots in Figure 3.8 represent combinations not covered by any gradsects. They illustrate the inevitable trade-off between completeness of coverage and resources available for the survey. It was decided that four gradsects was the maximum that could be sampled with available resources and those four were positioned to cover the maximum

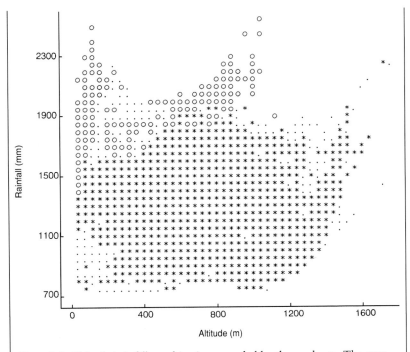

Figure 3.8. Altitude/rainfall combinations sampled by the gradsects. The stars represent combination sampled by the three southern gradsects and the open circles those combinations sampled by the addition of the fourth, most northerly, gradsect. The small dots are altitude/rainfall combinations not captured by the gradsects. (From Austin & Heyligers (1989). Reproduced with permission from Elsevier.)

number of rainfall/altitude combinations. Of the 43 combinations not covered, 18, nearly half, were represented by only one or two gridpoints. Thus, Austin and Heyligers were satisfied that the gradsects adequately represented the study area.

Sampling strategy

The next step was to develop rules for sampling within gradsects. The three major variables, rock type, rainfall and altitude, were divided into the classes shown in Table 3.1. The altitude and rainfall class intervals were based on previous experience with surveys and data analysis in similar environments (e.g., Austin *et al.* 1990).

This subdivision produced $9 \times 7 \times 8 = 504$ possible combinations or environmental cells. In fact, only 215 occurred within the study

Table 3.1 *Classes of environmental factors and their descriptions or ranges used to derive 215 environmental cells for field sampling.[a] There were nine rock types, seven altitude classes and eight rainfall classes*

Rock Type		Altitude		Average annual rainfall	
Class	Description	Class	Range	Class	Range
1	Quaternary alluvial, paludal and estuarine deposits. Mainly sands silts and mud	1	0–250 m	1	700–900 mm
2	Quaternary quartzose beach and dune sands	2	251–500 m	2	901–1100 mm
3	Basalts, tuffs and minor dolerite	3	501–750 m	3	1101–1300 mm
4	Adamellites and other "basic" granites	4	751–1000 m	4	1301–1500 mm
5	"Acid" granites	5	1001–1250 m	5	1501–1700 mm
6	Serpentine, ultrabasic intrusives	6	1251–1500 m	6	1701–1900 mm
7	Limestones	7	>1500 m	7	1901–2100 mm
8	Sedimentary rocks, mainly coarse grained e.g., conglomerates, greywackes, sandstones			8	>2100 mm
9	Sedimentary rocks, mainly fine-grained e.g. slates, phyllites, mudstones, argillites				

[a] Austin & Heyligers (1989). Reproduced with permission from Elsevier.

area, due mainly to the localized distribution of rock types. Sampling bias should be minimized by allowing each grid point in each environmental cell to have an equal, or at least known, probability of being sampled. This is impossible in practice because access to many grid points would require days of walking or a helicopter, both unacceptable costs. Instead, Austin and Heyligers adopted the following set of rules, which ensured consistency of sampling and, being explicit, provided an opportunity for the degree of bias to be determined.

Rule 1: Restrict sampling to within 0.5 km of tracks and fire trails accessible to vehicles.

Rule 2: Divide the gradsects into segments to allow for geographical replication where possible. (The short northern gradsect was not divided into segments.)

Rule 3: In each segment, the number of samples per environmental cell was determined as follows: (a) one grid point per cell: no sample; (b) 2, 3, or 4 gridpoints per cell: one sample if a site can be reached without undue demand on travelling time, e.g., en route to another sample site; (c) between 5 and 50 gridpoints per cell: one sample; (d) between 51 and 100 gridpoints per cell: two samples; (e) more than 100 gridpoints per cell: three samples.

Rule 4: If two or three samples are required they should be geographically separate if possible.

Locating sample quadrats

The next step was to determine the precise location of quadrats at each sample site. This had to be done in the field, but rules for quadrat location were specified before fieldwork commenced to ensure consistency and to minimize bias. These rules were also designed to provide another level of environmental stratification at the local site scale.

Rule 5: Each sample should consist of five quadrats to cover local environmental variation due to aspect and topographic position. They should be selected in the field from the following: a crest, an exposed north facing slope, an intermediate slope, either east or west facing, a south facing slope, a lower slope, a gully or a valley flat. Within the area delineated by the environmental cell, they could be located wherever the time and effort required to reach them is minimized. Where possible, this is to be within $1 \, km^2$ in order to minimize the risk of confounding the sample with previously unrecognized environmental or floristic variation. Figure 3.9 illustrates quadrat selection at the sample site level.

Rule 6: Each quadrat should measure $50 \times 20 \, m$ and be laid out with the long axis along a crest, gully or valley flat, or following the contour of a slope.

Rule 7: All canopy species, whether present as mature trees or saplings, should be listed, and the three most common species ranked according to abundance.

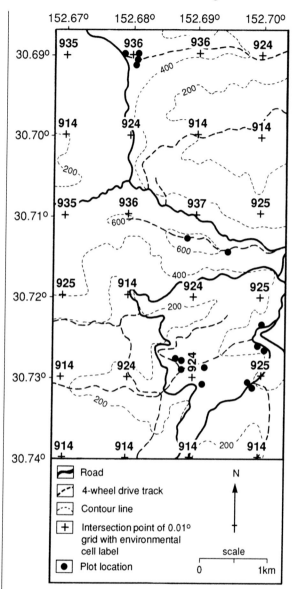

Figure 3.9. The locations of field-sample quadrats plotted on a topographic map. Clusters of five quadrats represent a sample site. The quadrats sample different topographic positions within the sample sites. See text for full explanation. (From Austin & Heyligers (1991). © CSIRO Australia 1991.)

Table 3.2 *Sampling results for the nine rock types*[a]

	Numbers and relative frequencies of grid intersection points					
	In study area		In gradsects		Samples obtained	
Rock type	Number	%	Number	%	Number	%
1	1 392	7.39	286	4.86	14	5.86
2	147	0.78	29	0.49	3	1.26
3	1 090	5.79	423	7.18	23	9.62
4	712	3.78	422	7.16	22	9.20
5	539	2.86	148	2.51	9	3.77
6	114	0.61	37	0.63	2	0.86
7	85	0.45	32	0.54	4	1.67
8	3 842	18.50	911	15.47	38	15.91
9	11 265	59.84	3 602	61.16	124	51.89

[a] Austin & Heyligers (1989). Reproduced with permission from Elsevier.

The survey recorded a total of 261 species in 909 quadrats. The New South Wales National Parks and Wildlife Service provided data from a further 116 quadrats in rugged inaccessible country, which added only a further two species. Thus the total number of quadrats available was 1025. Each was 0.1 ha in size, so the total area sampled was just over 1 km^2 in a region of approximately 20 000 km^2. Some methods for analyzing these survey data to derive communities of co-occurring species and predict wider distribution patterns from the point data obtained from the quadrats are described in Chapter 4.

Austin and Heyligers evaluated the representativeness of the survey by comparing the number of grid points in the study area with the number in gradsects, with, in turn, the number actually sampled for each rock type (Table 3.2). They concluded that the gradsects succeeded in capturing most of the variability in the study area, though more successfully for the more widespread rock types. There is further evidence of this success. From the 909 quadrats of the survey, 261 species were recorded. The addition of a further 116 quadrats from the New South Wales National Parks and Wildlife Service, which were added for the analysis described in Chapter 4, contributed only two more species.

Further examples of vegetation survey design, based on continuum theory and using environmental stratification, can be found in Austin *et al.* (2000) and Cawsey *et al.* (2002).

Example 3.3   Biogeographical survey of the Nullarbor District, Western and South Australia

In 1983, the Australian National Parks and Wildlife Service and agencies of the South Australian and Western Australian Governments initiated a collaborative biological survey of the Nullarbor district (McKenzie & Robinson 1987; McKenzie *et al.* 1989, 1991).

This is a warm-to-hot arid area of approximately 220 000 km² containing one of the largest continuous karst formations in the world. For comparison, the United Kingdom covers approximately 241 000 km². The most conspicuous karst phenomena are solution dolines, which form shallow circular depressions or gently undulating rectilinear ridges and corridors. These structures are several hundred meters across and topographic relief seldom exceeds 2 m (Curry 1987). Elevation in the north-west is 240 m sloping gently southward at a gradient of 1:200 and even more gently eastward at a gradient of 1:10 000 (Lowry & Jennings, 1974). The plain ends abruptly in a coastal marine escarpment. A narrow sandy plain has developed below this escarpment in the central part of the coastline but elsewhere the escarpment meets the sea (Fig. 3.10). Rainfall is seasonally uniform but spatially extremely variable. Mean annual rainfall at Forrest, more or less in the center of the district, is 187 mm. A low open shrubland, or desert steppe, occupies the central part of the district, hence the name Nullarbor, Latin for no trees. Low open arid woodlands occupy the remainder except in the south along the coast where woodlands and mallee (multi-stemmed *Eucalyptus* shrubs) both occur. In places, small areas of halophytic shrublands are found along occluded drainage lines. Pastoral leases are scattered in the south, but do not extend north of 30° south latitude because of insufficient water (Plates 15 and 16).

The main purpose of the survey was to provide an adequate database for the design of a conservation area network. This database should be sufficient to ensure that the conservation area network, "should encompass 1) most, if not all, of the species native to the region because these are the building blocks of the region's communities; and 2) not only the individual native species, but various alternative combinations

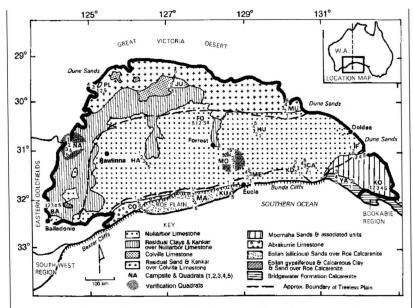

Figure 3.10. The Nullarbor region showing surficial geology, topographic features, place names, campsite locations and quadrat locations. (From McKenzie *et al.* (1989). Reproduced with permission from Elsevier.)

of species that form the naturally occurring assemblages, thereby sustaining as much biological complexity as possible" (McKenzie *et al.* 1989). Two important subsidiary purposes were to provide a base-line for monitoring the success or failure of conservation management practices and to evaluate a survey design based on environmental stratification for large remote arid regions.

Survey design

The vast size of the Nullarbor region and the limited resources available for the survey were a severe constraint on the survey design. There were four teams consisting of four biologists and their support staff to record plants, mammals, birds and reptiles.

The time taken had to be sufficiently brief to justify the assumption, necessary for the analysis and evaluation of the faunal data, that all observations were recorded simultaneously. Because of seasonal variations it was deemed necessary to conduct the survey twice, once in autumn and again in spring. The design proceeded in the following steps:

(1) Stratify the region using available maps of surficial geology, and broad vegetation types. For the Western Australian part of the region, broad vegetation maps were available (Beard 1975). For the South Australian part, aerial reconnaissance was necessary;
(2) Locate 16 campsites (four per team) representative of the range of vegetation and geological types across the region, and for ease of access;
(3) Locate five sample sites within 20 km radius of each campsite, to sample the array of local vegetation/geology combinations;
(4) Where possible, replicate these combinations at different campsites.

A brief reconnaissance field trip and iteration between steps 2, 3 and 4 was necessary before the final locations of campsites and sample sites could be determined (Fig. 3.10), but the practical constraints of limited resources and access requirements did not compromise the principles of environmental stratification and replication wherever possible.

Sampling strategy

A 4 km^2 quadrat was set out at each sample site. This large size was considered necessary because of the low standing biomass of the Nullarbor region. Reptiles and mammals were sampled using lines of pitfall traps and searches during the day and at night. Birds were censused between dawn and 10:00 each day, and all vascular plant species were recorded. Specific details of sampling methods can be found in McKenzie and Robinson (1987).

Summary of results

A total of 506 species were recorded from 80 quadrats. They comprised 96 birds, 10 ground-dwelling mammals, 59 reptiles, 157 ephemeral plants and 184 perennial plants. In all, 0.15 % of the Nullarbor region was sampled during less than 3 % of a single year. McKenzie *et al.* (1989) compared these figures with records from all previous surveys and collecting expeditions on the Nullarbor (Table 3.3) and concluded that the survey adequately sampled the target groups. Chapter 4, Example 4.2 describes how this data set was analysed to derive suitable surrogates for conservation planning.

Further examples of biological survey design over large areas in remote locations using environmental stratification can be found in McKenzie *et al.* (1991) and Burbidge *et al.* (2000).

Table 3.3 *Percentage of birds, reptiles and plants known from all previous surveys and collecting expeditions on the Nullarbor, that were recorded during the survey*[a]

	Proportion of all previous records (%)
Birds	
Passerines	85
Non-passerines	55
Reptiles	
Lizards	94
Snakes & Goannas	61
Plants	
Perennials	95
Ephemerals	65

[a] McKenzie *et al.* (1989). Reproduced with permission from Elsevier.

3.3.5 Survey gap analysis

A very common reason for biological surveys is to fill gaps in the coverage of previous collections. For example, we may want to fill the gaps shown in Figure 3.2. However, the best strategy is not to go directly to geographical locations that lack field records. As discussed above and illustrated in the two previous examples, species distribution patterns are most accurately and reliably mapped in environmental space and then transferred to geographical space. So the process of designing a survey to fill gaps is to plot existing field records in environmental space, identify unsampled or under-sampled environments and target geographical locations that have those unsampled or under-sampled environments. In this way the entire range of environmental variation within a region is more likely to be sampled and subsequent modeling of species distribution patterns (see Chapter 4) can rely on interpolation within the range of the data. Poon and Margules (2004) followed this process in searching for new populations of two rare plants on Cape York Peninsula in north-eastern Australia. They stratified the region of interest using environmental variables, identified strata in which populations of these two species had been recorded previously and searched for new populations in similar environments in the same general location, and in the same strata in different geographical locations (see Example 4.3, Plate 20). In this way, limited resources of labor, time and funds were used most efficiently. They found

Example 3.4   Filling gaps in the coverage of field records of vertebrates across the Wet Tropics, Queensland, Australia (Plates 7 and 8)

Figure 3.11 shows the geographical locations of survey sites where reptiles, amphibians, birds and mammals have been sampled in the Wet Tropics of north-east Queensland up until the field season of 2002. This figure also shows these sample sites in an environmental space formed by solar radiation and temperature, as an example. There are clear gaps in the coverage of environmental space by these field sites. There are large gaps in the high solar radiation and high temperature spaces. There is also a large gap in the mid to higher temperature range and low to mid solar radiation space and smaller gaps at lower temperatures and lower solar radiation. Figure 3.12 shows the increased coverage of this environmental space achieved with field-sample sites selected for survey in the field season of 2003. Solar radiation and temperature extremes have now been sampled and other gaps have

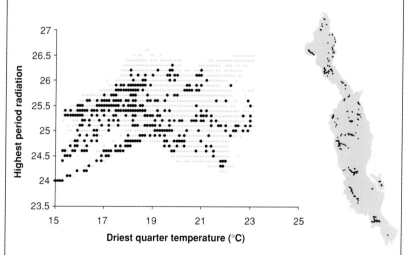

Figure 3.11. The locations of vertebrate field sites sampled up to and including the field season of 2002 in the Wet Tropics of Queensland (left) and plotted in environmental space formed by a solar radiation variable and a temperature variable (right). The grey and black dots combined represent all solar radiation by temperature combinations in the Wet Tropics and the black dots are those combinations representing sample sites. Note the gaps in coverage of environmental space, especially solar radiation extremes in the mid-temperature range and temperature extremes in the mid to high solar radiation range. (Data and map courtesy of Steve Williams and Karen Richardson.)

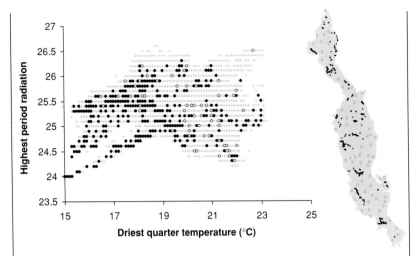

Figure 3.12. Vertebrate field sample sites selected for the field season of 2003 (open circles) in environmental space (left) and geographical space (right). These sites were chosen to, as far as available resources allowed, fill gaps in the coverage of environmental space. (Data and map courtesy of Steve Williams and Karen Richardson.)

been covered as far as possible, given the available resources of time, staff and funds. It is particularly important that the extremes of high and low solar radiation and temperature have all now been sampled. Spatial models to predict wider distribution patterns from these point records (see Chapter 4) will more likely represent interpolations within the range of the data than extrapolations beyond the range, and will therefore be more reliable.

one new population of one of the species and confirmed that both were indeed rare. In Chapter 2, a method called generalized dissimilarity modelling (GDM) (Ferrier 2002) was described, which predicts continuous patterns of turnover in species composition across whole regions. This same method can also be used to identify locations that are likely to be most different from existing survey areas, in terms of both biological and environmental composition.

3.4 Summary

Databases used in conservation planning are areas × features matrices. The areas are units of geographical space. They may be, for example, regular or irregular polygons, catchments or units of tenure. Features include the measures of biodiversity to be used in conservation planning. Inevitably, these are partial measures or surrogates for the whole of biodiversity, which is unmeasurable (Chapter 2). Biodiversity surrogates may be taxa (e.g., species), species assemblages and environmental classes or variables. In real planning applications, combinations of these attributes are commonly used.

Biological data in the form of records of the geographical locations of taxa may be available from previous collecting expeditions or surveys, or collected during new surveys. In designing new surveys it is crucial that the entire range of environmental variation is included in the stratification so that subsequent models of distribution patterns (Chapter 4) are interpolations within the range of the data, not extrapolations beyond it. Environmental data may be extracted, for example, from meteorological records and topographic maps, and from existing thematic maps of geology and soils. Many environmental variables can be modeled with, or extracted from, digital elevation models. Systematic conservation planning depends crucially on validly comparing sites with one another. Therefore, the data have to be at a consistent level of spatial resolution across the whole region of interest so that such comparisons can be made.

4 · *Data treatments*

Data that have been compiled from museums, herbariums, natural resource management agencies, conservation NGOs or researchers, or have been collected from new field surveys almost always require some form of analysis, or treatment, before they can be used for conservation planning. This is because raw data sets are not spatially consistent across planning regions and because raw data sets contain records of only a small proportion of biodiversity. Treatment can improve spatial consistency and can be used to derive higher-order surrogates that may represent more of biodiversity than species alone.

4.1   Conceptual framework

There are many procedures available for predicting feature (usually species) richness, for example by extrapolation (e.g., Hurlbert 1971; Margules *et al.* 1987; Palmer 1990; Prendergast *et al.* 1993; Colwell & Coddington 1994) and by smoothing richness across neighborhoods (e.g. Lawton *et al.* 1994; Williams & Gaston 1998) but they sacrifice information on feature identity. The richness of a place tells us nothing about which surrogates are there. Consequently, richness cannot be used to calculate complementarity and therefore cannot be used effectively to plan conservation area networks. Fortunately, a number of analytical procedures, which retain feature identity, are available to address the problems of spatial bias in biodiversity data sets. Figure 4.1, adapted from Margules *et al.* (1995), illustrates the different treatment pathways that can provide data in the form needed for conservation planning. Some of the more common and more promising analytical methods for use in these treatment pathways are discussed below.

Pathway 1. Records of the field locations of species either represent a comprehensive and unbiased spatial sample or are deemed sufficient. This may happen, for example, when the targets of conservation

80 · Data treatments

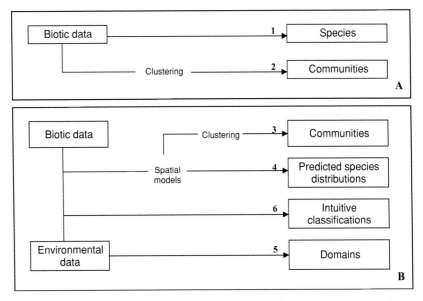

Figure 4.1. Six pathways commonly used to treat data for systematic conservation planning. The pathways in box A treat only biotic data. The pathways in box B utilize both biotic and environmental data. See text for more detail. (Redrawn from Margules *et al.* 1995.)

action are endangered and threatened species, the locations of which are already known or can be extracted from museum and herbarium records (Sarakinos *et al.* 2001; Conservation International 2004). No further analysis is required. Raw species data are used to select priority areas (e.g., Margules *et al.* 1988; Chapter 8 [Section 8.2]).

Pathway 2. Field locations of species are subject to multi-variate pattern analysis, commonly clustering, to form assemblages of co-occurring species (often called communities). Species assemblages are used to select priority areas (e.g., Margules & Nicholls 1987; McKenzie *et al.* 1989; Chapter 8 [Section 8.1]).

Pathway 3. Spatial models of wider distribution patterns are constructed from the field records of species and the locations of predicted distributions are clustered to form assemblages. Species assemblages are used to select priority areas. This potential pathway is included for completeness, but we could find no examples of this one being followed.

Pathway 4. Spatial models of wider distribution patterns are constructed from the field records of species. The predicted distribution patterns are used to select priority areas (e.g., Margules & Stein 1989; Austin *et al.* 2000).

Pathway 5. Environmental data are clustered or ordinated and either the classes (environmental domains) or the ordination pattern is used to select priority areas, usually in combination with assemblages or species (e.g., Belbin 1993; Faith et al. 2001b; Chapter 8 [Section 8.5]).

Pathway 6. Intuitive classifications of both biological and environmental data give rise to maps of habitat types (assemblages or communities) and these habitat types are used to select priority areas (e.g., Pressey & Nicholls 1989; Chapter 8 [Section 8.4]). Eco-regions are broader-scale examples of these intuitive classifications. The problem with them is that they are not repeatable and we have no idea of their internal spatial heterogeneity.

In many applications of conservation planning, more than one treatment pathway is used. This happens, for example, when vegetation types, or habitat types and species are both used (see Chapter 8 [Sections 8.4 & 8.5]).

There is theoretical support for the use of environmental variables to model wider spatial distribution patterns from the point records that field samples represent, which can be summarized as follows. Each species has a unique distribution within environmental space determined by its genetic constitution and physiological requirements. This distribution is, in turn, constrained by ecological interactions with other species. This is the concept of the niche (Hutchinson 1958). (Plant ecologists use the term "individualistic continuum" for essentially the same concept, see Austin [1985] and Chapter 3 [Section 3.3.2].) Species respond individualistically in, say, abundance or frequency, to resource gradients, and that response is constrained by interactions with other species. The implications are threefold: (1) each species occupies a unique niche not necessarily predictable from that of other species (i.e., there is little overlap in environmental space between many species, although much overlap between some); (2) related to the previous point, species distribution patterns are most accurately measured in multi-dimensional environmental space and only then translated to geographical space; and (3), the resultant spatial pattern shows high or dense populations in scattered locations representing the most favorable habitat (or mix of environmental variables) and lower, more sparse populations in areas of less-favorable habitat.

Thus, the geographical distribution patterns of species can be linked to variation in the environment. Whittaker (e.g., 1954, 1960), Perring (e.g., 1958, 1959) and Austin et al. (1984, 1990), among others, have provided empirical support. Nix (1986) has argued that, for many purposes, including estimating the spatial distribution patterns of species, complete niche specification is unnecessary and that in most cases, five regimes, namely solar radiation, temperature, moisture, mineral nutrients and other components of the biota, are sufficient.

Austin *et al.* (1984, 1990), Margules and Nicholls (1987), Nicholls (1989, 1991), Austin and Meyers (1996), Austin (2002) and Cawsey *et al.* (2002) have successfully shown how environmental variables such as rainfall, temperature, lithological substrate and solar radiation, among others, can be used to model the wider distribution patterns of trees and plant communities. Braithwaite *et al.* (1983, 1984, 1989) provide some empirical support for a relationship between animal species and environmental variables by establishing the link between levels of tree foliar nutrients and arboreal mammal abundance and bird species richness. Levels of foliar nutrients were linked, in turn, to the lithological substrate. Wessels *et al.* (1999) demonstrated a relationship between land facets (landscape units with uniform slope, soil and hydrological conditions) and communities of birds and dung beetles. Scott *et al.* (2002) provide a recent compendium and overview of the numerous methods now available for estimating species distribution patterns by linking field records with environmental variables.

This has led to the development of methods relating environmental variables to species presences, which can be statistical (e.g., regression), heuristic or built using computer-learning algorithms. Recently, these methods have come to be called bioclimatic envelop models (BEM) or niche models (Sections 4.3, 4.4 & 4.5 below). Problems may arise with these models when organisms are restricted to just part of their potentially suitable habitat by barriers to dispersal or low dispersal abilities, for example, species with relict distributions or vicariant species occupying similar, but geographically isolated, environments. Spatial distribution patterns in the hyperdiverse Fynbos of the Cape region of South Africa are a case in point (Cowling 1992; see Chapter 8 [Section 8.4] for a planning case study in the Cape Floristic Region). Species may also be restricted to sub-optimal habitat by practices such as land clearing, grazing and altered fire regimes. It is usually difficult to measure the impact of such factors on distributional patterns and incorporate them into spatial models. However, the problem can be alleviated somewhat by including geography as a factor in the models. They also ignore the effects on distributional limits of biotic interactions such as predation, competition and mutualism. Useful critiques of niche models, focusing on predicting how species distributions might change with climate change, can be found in Pearson and Dawson (2003) and Hampe (2004). However, for predicting wider distribution patterns from the point samples that field collections represent, they remain the most reliable suite of available techniques. Elith *et al.* (2006) have recently provided a comprehensive comparative survey of these methods.

4.2 Multi-variate pattern analysis

Pattern analysis is a form of exploratory data analysis where the objective is the detection of pattern or structure in the data. It is especially useful for sparse data sets. Many data sets that might be used for biodiversity priority setting are sparse in their coverage. The total number of records for many species may be too low for individual models of distribution patterns to be constructed with confidence. Clustering and ordination are the two most common methods of pattern analysis. In ecology and conservation biology, clustering is used to delineate groups of objects, usually regular polygons such as grid cells or quadrats used for collecting and storing field records, which have common features. If the features are species and the polygons are clustered according to similarity in species composition, then the resultant groups or clusters can be thought of as communities or species assemblages. If the features are environmental variables then the groups are called environmental classes or domains. Ordination is used to detect trends in similarity or difference in attribute composition between polygons.

Example 4.1 Environmental domains and vegetation types in Papua New Guinea (Plates 3 and 4)

Environmental domains

Environmental domains for PNG were derived from the data set described in Chapter 3 (Section 3.2) by clustering the 387 109 km^2 grid cells in the DEM. They were classified according to similarity in composition of the climate, terrain, landform and lithology attributes listed in Table 4.1. The bioclimatic parameters in Table 4.1 are biologically meaningful combinations of monthly mean climate values (Nix 1986; Hutchinson *et al.* 1996). Regular grids of these parameters were calculated by the BIOCLIM program in the ANUCLIM package (Hutchinson *et al.* 1996) by combining climate surfaces, which are functions of latitude, longitude and elevation, with the 0.01° DEM. Three of the parameters, elevation percentile, elevation difference and

---

Table 4.1 *Environment attributes recorded for each 1 km² grid cell and used to generate the environmental domains for the PNG conservation planning study (Example 4.1)*[a]

Attribute number	Description and units
Attribute group 1. Selected bioclimatic parameters	
1	Annual mean temperature (°C)
2	Temperature seasonality (CoV%)
3	Mean temperature hottest month (°C)
4	Mean temperature coldest month (°C)
5	Mean temperature wettest quarter (°C)
6	Mean temperature driest quarter (°C)
7	Mean temperature hottest quarter (°C)
8	Mean temperature coldest quarter (°C)
9	Annual mean precipitation (sq. root, mm)
10	Precipitation seasonality (CoV%)
11	Wettest quarter mean precipitation (sq. root, mm)
12	Driest quarter mean precipitation (sq. root, mm)
13	Annual mean radiation (MJ/m²)
14	Radiation seasonality (CoV%)
15	Wettest quarter mean radiation (MJ/m²)
16	Driest quarter mean radiation (MJ/m²)
Attribute group 2. Terrain and landform parameters	
17	Slope (degrees)
18	Curvature (m)
19	Relief (elevation range four nearest neighbors, m)
20	Relief 5k (elevation range 5 km radius, m)
21	Elevation percentile (elevation rank in 5 km radius, m)
22	Elevation difference (diff. from mean elevation in 5 km radius, m)
46	Area of mangroves (in grid cell)
47	Area of beach ridges (in grid cell)
48	Area of swamps (in grid cell)
49	Area of volcanic domes (in grid cell)
Attribute group 3. Lithological classes	
23	Fine-grained sediments
24	Coarse-grained sediments
25	Mixed or undifferentiated sediments
26	Mixed sediments or limestone
27	Limestone
28	High-grade metamorphics
29	Low-grade metamorphics
30	Mixed or undifferentiated metamorphics
31	Mixed sediments and volcanics

Table 4.1 (*cont.*)

Attribute number	Description and units
32	Basic to intermediate volcanics
33	Intermediate to acid volcanics
34	Mixed or undifferentiated volcanics
35	Mixed volcanics and limestone
36	Acid to intermediate volcanics
37	Basic igneous rocks
38	Ultrabasic rocks
39	Pleistocene sediments
40	Alluvial deposits
41	Marine sand
42	Estuarine deposits
43	Volcano–alluvial deposits
44	Pyroclastics
45	Overlain with recent ash

[a] Faith *et al.* (2001b). Reproduced with permission from Surrey Beatty & Sons Pty Ltd.

relief were calculated using in-house software (Centre for Resource and Environmental Studies, ANU). Other terrain variables were computed using ARC/INFO GRID. Six landform classes from the 40 contained in PNGRIS (Papua New Guinea Resource Information System, Chapter 3 [Section 3.2, Example 3.1]) were selected as landform features in addition to the terrain variables. Merging the three swamp classes reduced these to four. They are numbered 46–9 in Table 4.1. The lithological classes were generalized from the 900 digitized classes into the 23 classes recognized in PNGRIS, by Henry Nix and Janet Stein.

Clustering occurred in two stages. The first stage used a non-hierarchical procedure in the PATN software package called ALOC (Belbin 1987). It was necessary to do this because the data set was too large to run a hierarchical classification from scratch. In this first stage 2000 clusters were recognized. Subsequently, these 2000 were clustered using an agglomerative hierarchical procedure, also from the PATN software package, to derive a dendrogram, or classification tree, from 2000 up to one. This dendrogram is a representation of the environmental heterogeneity of PNG and could be cut at any level.

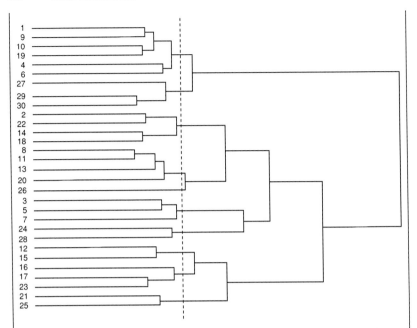

Figure 4.2. Part of the dendrogram from the classification of environmental domains in Papua New Guinea showing a line at the ten group level, which was used as a physio-climatic classification and intersected with 208 vegetation types to produce the final 564 vegetation types. (Redrawn from Nix *et al.* (2000).)

It would have been possible to recognize, for example, 500 environmental domains or 1000 or 1500 or, indeed, any number between one and 2000. In order to choose a number of domains, the current (1997) widely accepted target of 10% was adopted but implemented in a different way. Faith *et al.* (2001b) asked "How many domains (how far down the classification dendrogram), when combined with the vegetation types (described below) could be represented in any 10% of the country of PNG?" The answer is 608. As described in Chapter 8 (Section 8.5), the representation of these 608 domains became part of the conservation goal for PNG. Figure 4.2 shows a part of this dendrogram, from the 30 group level on up to one. Plate 1 is a map of the 608 environmental domains.

Vegetation types

There are 642 vegetation types on the PNG forest inventory map (Chapter 3). Some represent open water and urban areas and these were deleted. Others represent degrees of disturbance. For example, type Wsw is swamp woodland and Wsw8 is swamp woodland 80% undisturbed. Any type that was 70% or more undisturbed was merged with the wholly undisturbed type. Any type that was 60% or less undisturbed (40% or more disturbed) was regarded as too disturbed and was deleted. Types that were combinations of two or more original types, but were only distinguished from one another on the basis that in the first type, one was dominant and in the second, the other was dominant, were merged. For example, Hm/Wsw medium-crowned forest dominant over swamp woodland, was merged with Wsw/Hm, swamp woodland dominant over medium-crowned forest. Altogether, these mergers resulted in a reduction in vegetation types from the original 642 to 208. These 208 types are based mainly on vegetation structure and it seems certain that swamp woodland in the north-west of the country, for example, will contain many different species from swamp woodland in the south-east of the country. Because of this, the environmental domain classification was cut at the ten group level to produce ten physio-climatic zones (Fig. 4.2). The intersection of the 208 vegetation types with these ten zones resulted in 564 combinations and these were used as vegetation types to identify priority areas.

Example 4.2   Species assemblages in the Nullarbor region (Plates 15 and 16)

The Nullarbor region survey was described in Chapter 3 (Section 3.3). For the treatment of the survey data, it was deemed unrealistic to try to predict the distribution patterns of species recorded in 80 quadrats totalling $320 \, km^2$ in an area of $220\,000 \, km^2$. For many species, there were not enough observations to build spatial models with sufficient confidence to make predictions about their wider distributions and the data are subject to the limitation that a failure to record a species does not necessarily mean it was not there, especially when the quadrat that had to be searched was $4 \, km^2$. An alternative was adopted. Species were classified into assemblages according to the quadrats they shared in common and the assemblages were used as biodiversity surrogates to identify priority areas.

The survey recorded 506 species including birds, small mammals, reptiles, ephemeral plants and perennial plants. As it turned out, many of these could not be used in the analysis because the quadrats were searched too briefly to be sure that no record meant they were truly absent. In addition, wide-ranging species such as snakes and goannas

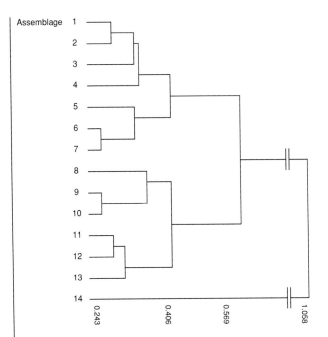

Figure 4.3. The dendrogram, at the 14 group level, of the classification of species assemblages in the Nullarbor Region. (Redrawn from McKenzie *et al.* (1989). Reproduced with permission from Elsevier.)

A. Four assemblages containing arid to semi-arid species of heavy, fine-textured soils, found in the eastern, western and central parts of the region:

    Assemblage 1 – widespread woodland species;

    Assemblage 8 – ubiquitous species of woodlands, shrublands and grasslands;

    Assemblage 9 – species confined to shrublands of the treeless plain, very open low woodlands and clearings within woodlands;

    Assemblage 10 – species confined to saltpans to the south, east and west of the treeless plain.

B. Arid species of well-drained desert sands and deep loams, found in the north of the region:

    Assemblage 11 – open woodland species of the north and north-western periphery;

    Assemblage 12 – open woodland species of the north-eastern periphery;

    Assemblage 13 – woodland species of the north-western and western periphery.

C. Mesic to semi-arid species found in sub-coastal and coastal areas of the Nullarbor region:

    Assemblage 2 – species of the most mesic woodlands and mallees;

    Assemblage 3 – species of the mallee scrubs of the Hampton Range;

    Assemblage 4 – species of semi-arid woodlands in the south-east and south-west extremities of the region;

    Assemblage 5 – ubiquitous species of woodlands and mallee;

    Assemblage 6 – species of unconsolidated beach dunes;

    Assemblage 7 – two species only, with a disjunct distribution;

    Assemblage 14 – species of consolidated marine and dunes on the Roc Plain.

Table 4.2 *Discrepancy between expected and observed richness for each assemblage at the ten new quadrats sampled*[ab]

	Assemblage													
Quadrat	1	2	3	4	5	6	7	8	9	10	11	12	13	14
NA1	0.45	B	A	A	A	A	–	0.23	0.30	A	0.30	A	A	A
NA2	0.21	A	A	B	A	A	–	0.25	A	A	B	0.24	0.21	A
NA3	A	A	A	B	A	A	–	A	A	A	A	B	A	A
NA4	B	A	A	A	A	A	–	A	A	A	A	B	B	A
NA5	A	A	A	A	A	A	–	A	A	A	A	B	A	A
MO1	A	A	A	A	A	A	–	A	0.29	A	A	A	A	A
MO2	A	A	A	A	A	A	–	A	0.23	A	A	A	A	A
MO3	A	A	B	A	A	A	–	A	0.31	A	A	A	A	A
MO4	A	A	B	A	A	A	–	A	0.23	A	A	A	A	A
MO5	A	A	B	A	A	A	–	A	0.21	A	A	A	A	A

[a] McKenzie et al. (1989). Reproduced with permission from Elsevier.
[b] The figures are the proportion of assemblage observed – proportion expected. Discrepancies in the proportions that were less than 0.20 are categorized as: A (< 0.10) or B (0.10–0.19). Assemblage 7 was too small for interpolation; the two species comprising this assemblage were neither observed nor expected in the vicinity of the MO and NA quadrats.

and certain non-passerine birds such as raptors, the Australian bustard and emu would not be continuously resident on quadrats. Communities of ephemeral plants experience rapid turnover in species composition (Grubb et al. 1982) and more intensive sampling would have been needed to provide reliable data on this group. Finally, 63 species occurred on only one quadrat. Most of these quadrats were near the boundary of the region and the species were more common in adjacent areas. After the elimination of all of these, there were 286 species; 152 perennial plants, 88 birds, 47 reptiles and 9 mammals.

Analysis

The two-step method (Belbin 1980; Austin & Belbin 1982) was used to measure the quantitative relationship between each pair of species. The first step generates a matrix with zeros in the diagonal and values ranging from zero (identical) to one (no overlap of species) in the off-diagonal positions. The second step accepts this matrix and determines the difference between the rows by subtraction. This procedure provides a more robust method of assessing similarity than a simple pair-wise comparison. It also provides a measure of association between species that may not co-occur, but may be adjacent on an environmental gradient (McKenzie *et al.* 1989). An agglomerative hierarchical-clustering method called unweighted pair group arithmetic averaging, or UPGMA (Sneath & Sokal 1973; Belbin, 1984) was used to group species into assemblages.

In theory, there could be any number of groups less than or equal to 285. But after inspection of the classification and group composition at various points along the dendrogram, only 14 were considered to be ecologically meaningful (Figure 4.3). The result was a matrix of 80 quadrats by 14 assemblages. Each value in the matrix referred to the proportion of the species in an assemblage that was present in each quadrat.

A partial test of the data and models embodied in the assemblages was carried out when the region was re-visited two years later. Ten new quadrats were sampled, five in the west and five near the center of the region (campsites NA and MO respectively in Figure 3.10). Table 4.2 compares the actual richness with that expected from the assemblage models at each of the ten quadrats. In all but 13 of the 140 cases, the predictions were better than 80% correct.

4.3   Heuristic models

One well-established and commonly used spatial-modeling technique is BIOCLIM (Nix 1986; Busby 1991). It is based on the idea that long-term monthly mean meteorological data can be used to represent energy and water balances, which in turn are good predictors of species distribution patterns. BIOCLIM works by generating climatic indices for the locations of field records of species and then finding other locations with the same or similar indices, which thus map potential distribution patterns. The first step is to model monthly mean minimum and maximum temperature and precipitation for the entire region of interest. As in the case of the derivation of the base data for the domain classification in Papua New Guinea described above, this is done by fitting a function of three independent variables, latitude, longitude and elevation, to climatic values measured at climate stations. The function is a trivariate spline, with spatial interpolation made by use of a digital elevation model (DEM) (Hutchinson 1991, 1995). The next step is to derive climatic indices from these surfaces. The current published version of BIOCLIM contains 16 indices (Busby 1991). Other versions have been developed using a larger number of indices, including solar radiation, as was done for the PNG domain classification (Table 4.1). Climate indices are generated for grid points throughout the study area. The resolution of these grid points varies depending on the resolution of the DEM. The climatic indices for sites at which a species (or other attribute such as vegetation type) has been recorded can then be aggregated into a climatic profile for that species. The degree to which the climatic profile of a species and the indices at grid points match can be used to indicate different levels of climatic suitability for that species (Busby 1991; Elith & Burgman 2002). Grid points that fall outside the range of climates sampled for a species are considered unsuitable. Those that fall within the range but outside the extremes of the 5–95 percentiles are considered marginal. Those falling within the 10–90 percentiles are considered to have the most suitable climate for that species.

Because this method uses presence-only data, it is restricted to estimating the geographical range of a species. It cannot predict absences within that range. Large areas may be included as potential occurrence space due to the presence of a single extreme observation (Walker & Cocks 1991; Carpenter et al. 1993). Thus the method is limited to estimating potential distributions. There are no quantitative predictions or statistical tests and confidence limits. Example 4.3 illustrates the use of BIOCLIM.

On the other hand, presence-only data are by far the most common form of biological field records. BIOCLIM is a tool for adding value to existing data of this kind and reducing its inherent spatial bias. Published BIOCLIM applications include the spatial estimation of the distribution patterns of elapid snakes (Nix 1986), C3 and C4 grasses (Prendergast & Hattersley 1985), temperate rainforest tree species (Busby 1986), weeds (Panetta & Dodd 1987), *Eucalyptus* trees, rainforest vertebrates (Nix & Switzer 1991), rare plants (Elith & Burgman 2002) and the spatial pattern of damage to vegetation by the soil pathogen *Phytophthera cinnamomi* (Podger et al. 1990).

Another heuristic method is DOMAIN (Carpenter et al. 1993), which differs from BIOCLIM in that it measures the climatic similarity between candidate sites and the nearest (in climatic space) sites with records, using the Gower metric (Gower 1971). A continuous function is generated so that degrees of similarity can be selected (Austin et al. 1994).

HABITAT (Walker & Cocks 1991) is also an heuristic method. It can work with a variety of environmental data, including climatic indices

Example 4.3   BIOCLIM model for *Coix gasteenii* on Cape York Peninsula, Australia

In a recent example, Poon and Margules (2004) used BIOCLIM to identify areas that might contain new populations of the rare grass, *Coix gasteenii*, in a remote, relatively inaccessible part of Cape York Peninsula, Australia (Plate 20). Searching for new populations of rare species in the field, especially in remote locations with limited access, is expensive and time consuming. Poon and Margules used BIOCLIM to locate areas climatically similar to the known locations with populations of *C. gasteenii* (Figure 4.4). By restricting field searches to those locations they were able to use the time and resources they had available for field survey efficiently. They discovered one new population and confirmed that the species is indeed rare.

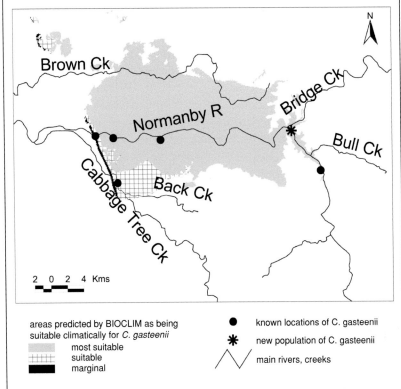

areas predicted by BIOCLIM as being suitable climatically for *C. gasteenii*

    most suitable

    suitable

    marginal

● known locations of *C. gasteenii*

✳ new population of *C. gasteenii*

∧∨ main rivers, creeks

Figure 4.4. Areas with climates similar to those at known locations of populations of *Coix gasteenii*, derived from BIOCLIM. (From Poon & Margules (2004). © 2004 Island Press.)

derived from BIOCLIM. In addition, it can use categorical attributes such as soil type and geological substrate. HABITAT is similar in concept to BIOCLIM in that it generates an "environmental envelope" from which potential distribution is determined. Unlike BIOCLIM, it treats the attributes as interdependent influences on potential distribution. It uses linear programming to describe the envelope occupied as a function of the linear combination of the set of attributes.

4.4 Regression models

If the data are in the presence–absence form, then an appropriate analytical method is statistical modeling using regression analysis to relate the presence or absence of a species to environmental predictor variables. There are numerous examples of both generalized linear modeling (GLM) and generalized additive modeling (GAM) for estimating species distribution patterns (e.g., Austin 1987, 2002, 2005; Austin et al. 1984, 1990, 1994; Nicholls 1991; Leathwick & Mitchell 1992; Pausas et al. 1997; Ferrier et al. 2002; Guisan et al. 2002). Though GLM and GAM should normally only be used with presence–absence data, they have been used with presence-only data by treating a random set of points with no presence as "pseudo-absences" (Ferrier & Watson 1997; Ferrier et al. 2002).

Example 4.4   A model of the distribution of *Eucalyptus radiata* in northern New South Wales

Nicholls (1989) constructed a GLM of the distribution of *Eucalyptus radiata* from the data collected in the survey of coastal hardwood forests described in Example 3.2 in Chapter 3, which is repeated here as an example. Other examples can be found in the references listed in the paragraph above. Figure 4.5 shows the geographical distribution of *E. radiata* recorded in the survey, and Figure 4.6 shows the environmental distribution of those field records in rainfall and temperature space. Table 4.3 lists the environmental variables that were available for use as predictors in the spatial model. Three of them are continuous and three are categorical. The data from the survey were presence–absence and so it was assumed that they were drawn from a binomial distribution, which suggests that a logit function linking the linear predictor to the response variables is appropriate. A forward stepwise procedure was used. Each variable was fitted alone and the variable with the

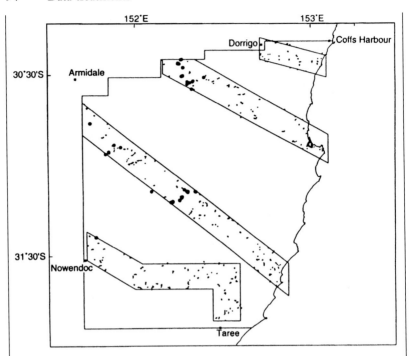

Figure 4.5. The geographical distribution of E. *radiata* recorded in the survey described in Example 3.2 in Chapter 3. (From Nicholls (1989). Reproduced with permission from Elsevier.)

greatest change in deviance was then added to the model and each variable fitted again, with the one associated with the greatest change in deviance added and so on until there were no significant changes in deviance. A quadratic response to the continuous variables was also tested as this shape of response is commonly observed in nature, for example, a species occurs with highest frequency in the middle of a rainfall gradient and low frequency at either end, where there is very high or very low rainfall.

Table 4.4 shows the change in deviance when each variable from Table 4.3 is fitted alone to the mean. The two variables, altitude and temperature, when fitted as linear or quadratic functions, account for the greatest change in deviance, with lithology next. The quadratic temperature function was selected because temperature is more biologically meaningful than altitude and as they are likely to be strongly correlated, altitude would be less significant when fitted to a model

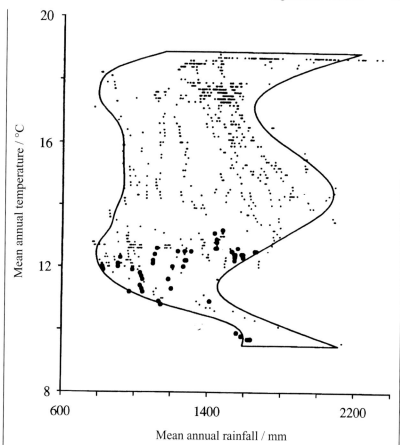

Figure 4.6. The location of field records of *E. radiata* in environmental space formed by mean annual temperature and mean annual rainfall. (From Nicholls (1989). Reproduced with permission from Elsevier.)

that already contains temperature, as proved to be the case. Table 4.5 shows the changes in deviance when each remaining variable was fitted to a model that already included temperature. Lithology accounted for the greatest change in deviance. When the remaining variables were fitted to a model with temperature plus lithology, the quadratic function in rainfall was the most significant variable. No other variables were significant at the 5 % level when added to a model containing temperature, rainfall and lithology. Table 4.6 shows the parameter estimates and standard errors for the *E. radiata* model after this fourth and final step.

Table 4.3 *Environmental variables available for modeling the spatial distribution of* Eucalyptus radiata[a]

Variable	Type	Range or number of levels
Altitude	Continuous	0–1750 m
Mean annual rainfall	Continuous	800–2300 mm
Mean annual temperature	Continuous	9.0–19.9 °C
Lithology	Categorical	9
Topography	Categorical	6
Exposure	Categorical	3

[a] Nicholls (1989). Reproduced with permission from Elsevier.

Table 4.4 *The first step in a forward stepwise procedure for fitting a model for* Eucalyptus radiata. *Each variable was fitted alone*[a]

Model	Residual deviance	d.f.	Change in deviance	d.f.
Mean	440.16	1024		
Altitude	267.22	1023	172.94	1
Altitude + altitude^2	257.54	1022	182.52	2
Rainfall	432.80	1023	7.36	1
Rainfall + rainfall^2	430.50	1022	9.66	2
Temperature	269.29	1023	170.87	1
Temperature + temperature^2	259.41	1022	180.75	2
Lithology (9)	350.50	1016	89.66	8
Topography (6)	432.97	1018	7.19	5
Exposure(3)	434.78	1022	5.38	2

[a] Nicholls (1989). Reproduced with permission from Elsevier.

Table 4.5 *The second step in the forward stepwise procedure, where each remaining variable was fitted alone to a model containing temperature plus temperature squared*[a]

Model	Residual deviance	d.f.	Residual deviance	d.f.
Temperature + temperature^2	259.41	1022		
Altitude	257.04	1021	2.37	1
Altitude + altitude^2	256.75	1020	2.66	2
Rainfall	253.00	1021	6.41	1
Rainfall + rainfall^2	249.82	1020	9.59	2
Lithology (9)	199.74	1014	59.67	8
Topography (6)	257.08	1017	2.33	5
Exposure (3)	256.72	1020	2.69	2

[a] Nicholls (1989). Reproduced with permission from Elsevier.

Table 4.6 *Coefficients and standard errors for the model of* E. radiata *after the fourth and final step[a]*

Variable	Parameter estimate	Standard error
Mean	−34.88	47.59
Temperature	4.396	6.166
Temperature^2	−0.2624	0.2557
Rainfall	0.0417	0.0108
Rainfall^2	−1.577e^{-5}	4.236e^{-6}
Lithology (1)	0.000	
Lithology (2)	2.358	91.02
Lithology (3)	−11.78	36.90
Lithology (4)	−6.305	36.91
Lithology (5)	−4.069	36.92
Lithology (6)	−10.19	78.17
Lithology (7)	−0.9083	36.73
Lithology (8)	−6.741	36.91
Lithology (9)	−8.222	36.91

[a] Nicholls (1989). Reproduced with permission from Elsevier.

Table 4.7 *Coefficients of a revised final model after removing lithological substrate classes 1, 2 and 6, and the* temperature^2 *term[a]*

Variable	Revised model	
	Parameter estimate	Standard error
Mean	−2.698	7.691
Temperature	−2.085	0.364
Temperature^2		
Rainfall	0.041	0.011
Rainfall^2	−1.55e^{-5}	4.20e^{-6}
Lithology (1)		
Lithology (2)		
Lithology (3)	−4.159	1.247
Lithology (4)	1.857	0.546
Lithology (5)	4.124	1.149
Lithology (6)		
Lithology (7)	5.885	1.763
Lithology (8)	1.481	0.556
Lithology (9)	0.000	0.000

[a] Adapted from Nicholls (1989). Reproduced with permission from Elsevier.

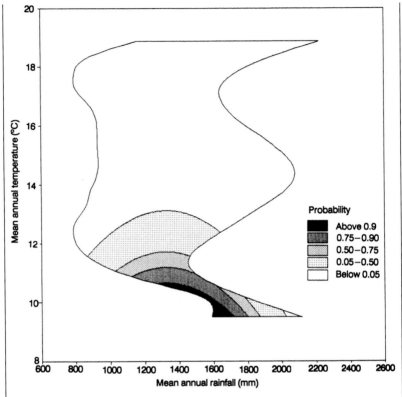

Figure 4.7. Predicted distribution of *E. radiata* in environmental space from coefficients of the final model in Table 4.7, on fine-grained sediments. (From Nicholls (1989). Reproduced with permission from Elsevier.)

There is a final twist in the tale of this model. Note that in Table 4.6 the standard errors of the parameter of the lithological substrate classes and temperature are all large relative to the parameter estimates themselves. Returning to the data, it was found that *E. radiata* was not recorded from lithological substrate classes 1, 2 and 6 (alluvium, coastal sands and serpentines) so the model was re-fitted with sites from these substrate classes omitted. Similarly, when the quadratic temperature term was removed and the associated change in deviance compared to the chi-squared distribution with one degree of freedom it was found to be not significant. This term was also omitted from the model. The re-fitted model then had only a linear term in temperature which was significant. Table 4.7 gives the coefficients for the revised final model. It was important to examine the model critically at each stage of the

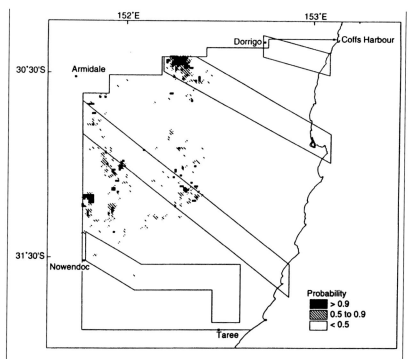

Figure 4.8. Predicted geographical distribution and probability of occurrence of *E. radiata.* (From Nicholls (1989). Reproduced with permission from Elsevier.)

procedure. Even when a final model with only significant terms was arrived at, it was still necessary to examine the model and return to the data for one more revision.

The predicted probability of occurrence of *E. radiata* was calculated by including the parameters from Table 4.7 in the following equation:

$$\mu = 2.698 + \text{Lithology}(i) + 0.041r - 1.54e^{-6}r^2 - 2.085t$$

where *r* and *t* are mean annual rainfall and temperature and lithology(*i*) is the parameter for the appropriate lithological substrate class. The estimate μ is the linear predictor and is transformed to a probability using an appropriate equation for binomial data (see Nicholls 1989). The probability of occurrence of *E. radiata* in all rainfall/temperature combinations was calculated separately for each lithological substrate class, and the resulting probability surfaces contoured. Figure 4.7 shows the predicted probability in rainfall and temperature space on fine-grained sediments and Figure 4.8 shows the predicted geographical distribution of *E. radiata* in all substrate classes across the entire study area.

4.5 Machine-learning methods

In recent years a wide variety of computational methods have been developed to model species' environmental preferences from a set of geographical occurrence points (longitude/latitude pairs). Like the heuristic methods described above, these methods are particularly useful when data are presence-only, which precludes the straightforward use of general purpose statistical techniques. Modeling occurs in an environmental space and models are then projected back to geographical space for interpretation and, in the context of systematic conservation planning, area prioritization. Typically models such as these are most reliable when the fundamental and realized niches of each species coincide. Care must be taken to ensure that there is temporal correspondence between the point occurrence data and the environmental layers, and also that the spatial resolution of the different data sets are compatible.

Neural networks (e.g., Hilbert & van den Muyzenberg 1999), genetic algorithms (e.g., Stockwell & Noble 1992; Stockwell & Peters 1999), and maximum entropy modeling (Phillips *et al.* 2006) are some of the computational methods currently in use for niche modeling. They are all machine-learning methods in which part of the data set is used to "train" the algorithm and the rest used to test the training. The Maxent method (a maximum entropy technique, as the name indicates) has been used with greater predictive success than competing methods (Phillips *et al.* 2004, 2006; Elith *et al.* 2006). Maxent is similar to standard statistical methods such as maximum-likelihood estimation. Maxent is potentially valuable for conservation planning because hundreds of species can be simultaneously modeled using freely downloadable software. An application is described in detail below (Example 4.5).

Maxent aims to estimate the unknown probability distribution of the occurrence of a species, given geo-referenced records of its presence. It is based on the assumption that the best such estimate is the probability distribution of a parameter known as the maximum entropy subject to the constraint that the expected value of each environmental variable under this estimated distribution is equal to its empirical average. (Formally, the entropy is the same function as the Shannon index of alpha diversity, with the probability of occurrence of a single species at different geographical locations replacing the relative frequency of each species in the Shannon index.) The output from Maxent consists of relative probabilities of occurrence for species in each cell of the geographical space.

Example 4.5   Potential geographic distributions for *Bradypus variegatus* and *Microryzomys minutus*

Phillips *et al.* (2006) modeled the distribution of the brown-throated three-toed sloth, *Bradypus variegatus*, and a small-bodied rodent, *Microryzomys minutus*, in South America. Figure 4.9 shows the geographical locations of records of occurrence of the two species. Three types of environmental variables were used: climate, elevation and potential vegetation. A $0.05° \times 0.05°$ pixel size yielding a 1212×1592 grid was used. Climatic variables were derived from data provided by the Intergovernmental Panel on Climate Change (IPCC: New *et al.* 1999). These data were at a resolution of $0.5° \times 0.5°$ and bilinear interpolation was used to resample the data. Twelve climatic variables were selected on the basis of their potential influence on the distribution of these two species: annual cloud cover, annual diurnal-temperature range, annual frost frequency, annual vapor pressure; January, April, July, October and annual precipitation; and minimum, maximum and

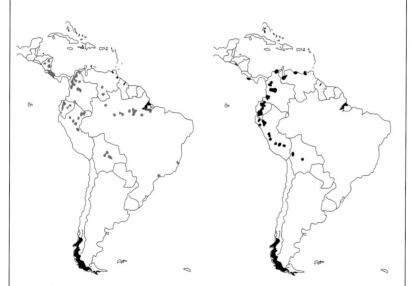

Figure 4.9. Occurrence records for *Bradypus variegatus* and *Microryzomys minutus*. Records for *Bradypus variegatus* (left) and *Microryzomys minutus* (right), as derived from vouchered museum specimens. (From Phillips *et al.* (2006). Reproduced with permission from Elsevier.)

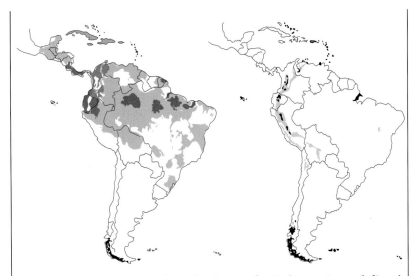

Figure 4.10. Potential geographical distributions for *Bradypus variegatus* (left) and *Microryzomys minutus* (right), as predicted by Maxent. Four shades of grey are used to indicate the strength of the predictions with darker shades indicating stronger predictions. (From Phillips *et al.* (2006). Reproduced with permission from Elsevier.)

mean annual temperature. In addition two other variables were used. An elevation variable was derived from a DEM (USG Survey 1998), with the resolution reduced from the original 1 km² by resampling. A habitat-type variable was used which divided South America into 15 major habitat types (Dinerstein *et al.* 1995).

For each species Phillips *et al.* made ten random partitions of the occurrence localities. Each partition randomly selected 70% of the occurrence data for training with the remaining 30% for testing. (For *Bradypus variegatus* there were 81 training and 35 test localities; for *Microryzomys minutus*, 61 training and 27 test localities.) The ten partitions permitted an assessment of the average performance of the algorithms. Maxent was used with two suites of environmental variables: all three types together, and then only climate and elevation (i.e., excluding habitat type). The predicted distributions, shown in Figure 4.10, passed statistical tests of commission and omission errors.

The main limitation of such computational algorithms is over-prediction. Typically, the predicted presence is larger than the observed presence. Therefore, before a predicted distribution is used for conservation planning purposes, a series of steps should be taken to remove areas known to be unlikely to be inhabited by a species. For example, suitable areas that are unlikely to be colonized due to contingent reasons (such as geographical barriers) should be excluded (Peterson *et al.* 1999; Anderson 2003). Suitable areas likely not to be inhabited due to ecological reasons (such as competition with related species) can similarly be excluded (Anderson *et al.* 2002). Remote-sensed data can be used to exclude anthropogenically or otherwise transformed areas that, even if presently occupied by a species, are likely to be unsuitable for their persistence into the future (Anderson & Martínez-Meyer 2004).

4.6 Summary

There are a number of options for the treatment of data sets to make them suitable for systematic conservation planning. One is to estimate the geographical distribution patterns of taxa, either intuitively, or by relating actual records of locations to environmental variables using predictive empirical models such as BIOCLIM (Nix 1986; Busby 1986, 1991), statistical models such as GLMs (e.g. Nicholls 1989, 1991) and GAMs (e.g. Austin & Meyers 1996) or machine-learning methods such as Maxent. Empirical models and machine-learning methods can be used with data of the presence-only kind, that is, there are records of where a taxon occurs, but it is not known whether a non-occurrence is a true absence or simply a result of the taxon not having been looked for there. Almost all museum and herbarium accessions, one of the most common and widely available sources of field records, are of this kind. Statistical models are appropriate for data of the presence–absence kind. That is, the absence of a species is the result of it having been looked for, but not found. Statistical models have the advantage that the relative significance of different variables can be readily assessed.

The other option is to classify the data into assemblages and map these assemblages. Classification can be done intuitively or with numerical methods, and mapping can be intuitive or it can utilize computer-based empirical or statistical modeling methods. Similarly, if the intention is to use environmental classes, classification can be intuitive or numerical, and mapping can be manual or computer based.

Whether it is biological, environmental or a combination of both, the end result is a data set that contains maps (on paper or in electronic form) of the chosen biodiversity surrogates. This is the data set that can then be used to identify biodiversity priority areas.

5 · *Conservation area networks*

Conservation area networks are sets of areas within a planning region, which should have priority for the allocation of biodiversity management resources. Identifying such priority areas and separating them from threatening processes is the purpose of systematic conservation planning. This chapter discusses the general conceptual and theoretical issues that are relevant to the design of conservation area networks. Chapters 6 and 7 discuss further refinements of conservation area networks incorporating ideas on persistence and vulnerability, socio-economic and other costs, and the spatial constraints imposed by land use history (including conservation areas that have already been established) and suitability for alternative uses. Chapter 8 describes detailed case histories of how these principles are used in practice for the selection of conservation area networks to represent biodiversity.

5.1 The role of conservation area networks

A *conservation area* is a place that is being managed at least partly to conserve the biota present in it. Conservation areas can range from vast national parks such as the 345 000 km² Great Barrier Reef Marine Park in Australia, and the 20 000 km² Kruger National Park in South Africa, to tiny sacred groves in the Western Ghats in India. Management options range from the exclusion of permanent human presence, as in many national parks and wilderness reserves throughout the world, to community-based conservation that may even include sustainable levels of harvesting of certain targeted species.

Traditionally, conservation areas have been thought of only as national parks, nature reserves or wilderness areas with no human use such as resource extraction. It has been assumed that human use is necessarily detrimental to the biota. While this is often the case because much of current human activity is ecologically unsustainable, it is a mistake to accept this assumption as a fact that does not require further empirical support.

Example 5.1 Communal conservancies in Namibia (Plates 9 and 10)

Communal conservancies in Namibia allow the practice of farming activities, mostly livestock grazing, in combination with wildlife use on a sustainable basis. Communal conservancies are formed when adjacent communities decide to pool resources in order to conserve and use wildlife sustainably, and the Namibian government sanctions and gazettes the areas involved. For this to happen, it must first be determined that the proposed conservancy is ecologically, socially and economically viable. If it is, then landowners elect a management committee, which draws up a constitution and defines the boundaries of the proposed conservancy. Once conditions of registration are met, conservancies obtain limited rights of ownership over some animals and use rights over others. Once conservancies are legally recognized, they can also apply for hunting and tourism rights. In 1998, the first communal conservancy, Nyea Nyea, was gazetted. By the end of 2003, 15 conservancies, occupying over 38 500 km^2, had been legally recognized and more than 30 were at various stages of establishment. Conservancies in Namibia have come to form a significant addition − a kind of second tier conservation function − to the protected area network. They thus play an important role in the protection of biodiversity. (For more details, see www.dea.met.gov.na/programmes/cbnrm/cbnrm.htm.)

For instance, the Keoladeo National Park in Rajasthan (India) is a 450 ha artificial wetland that is a premier bird habitat, but has traditionally also served as a grazing ground for cattle (Gadgil & Guha 1995). On the advice of Indian and US experts, grazing was banned in the early 1980s in an effort to promote bird diversity after the designation of the area as a national park in 1981. The ban devastated Keoladeo as a bird habitat. Paspalum grass and other opportunistic weeds, which had been kept in check by grazing, established a stranglehold on the wetland leading to declines in bird populations (Vijayan 1987). The ban is no longer enforced, though it has not formally been revoked, and anecdotal evidence suggests that bird diversity seems to have recovered (Lewis 2003).

Many traditional agro-ecosystems are effective in conserving elements of biodiversity as well as producing food (Halladay & Gilmour 1995;

Harvey *et al.* 2004). Old pastures managed without fertilizers in Norway have up to 45 plant species per m^2 compared to 27 per m^2 in pastures that are no longer grazed and 14 per m^2 in fertilized pastures (Götmark 1992). Areas that are set aside mainly to protect ecosystem services such as the delivery of clean water, or to protect recreational and scenic values, also protect biodiversity. Conservation areas today are managed under a variety of land tenures and operate within a great variety of social contexts. About half of the protected areas world-wide have significant human populations living within them (Kanowski *et al.* 1999). In India, for example, it is estimated that about 4 million people live within officially designated protected areas. In Papua Province of Indonesia (formally, Irian Jaya), on the western side of the island of New Guinea, between 6000 and 10 000 people live in the Lorentz World Heritage Area. It is now accepted that many species and habitats of value in Europe are found outside protected areas, and that future protection of such areas depends on development and planning strategies for agriculture, transport, urban development, etc., being integrated with biodiversity protection strategies and applied throughout the countryside in which people live and work (see, for example O'Riordan *et al.* 2002).

It has also traditionally been assumed that the best management is non-intrusive, that is, active human intervention is undesirable. However, as the example of Keoladeo shows, especially for small conservation areas in human-dominated landscapes, non-intrusive management is not a viable option. The best management option for a conservation area can only be determined by empirical research.

Conservation area networks should primarily be seen as holding operations, repositories of biota for evolution to work with in the future. Many, perhaps most, species will not be secure even in conservation areas, and many others do not need such areas for security. Conservation areas are components of landscapes, not something apart. For some species, conservation areas may act as refuges of optimal habitat in times of stress, for others they provide only suboptimal habitat and for yet others they may provide the only suitable habitat remaining, be it optimal, suboptimal or marginal.

The role of conservation area networks is, therefore, to encompass a true or "representative" sample of biological diversity, protect that sample from threatening processes and sustain it into the future. This may not be an achievable goal given our imperfect knowledge of biodiversity and the fact that there is an upper limit on the total area that will ever be allocated to conservation. But it is an end to work towards. Unfortunately, the practice of conservation planning, in particular, the identification of conservation areas, has not usually worked towards this goal. New conservation areas have often been located in places that do not add significantly to the representation of biodiversity. This is mainly because conservation areas compete with other economically valuable uses of land, which mostly win out in direct competition. As a result, such areas tend to be concentrated in places that are remote or too unproductive to be economically important (Pressey *et al.* 1996), which in turn means that the biodiversity of productive and economically important places is poorly represented, or not represented at all, in these networks. This is also due in part to the very diversity of goals that nature protection may address, such as wilderness values, scenic grandeur, etc. There is competition among proponents of protection for limited funds and the limited attention spans of decision-makers. While the protection of wilderness also protects elements of biodiversity, a focus on wilderness is a focus on remote, usually unproductive places at the expense of accessible, often economically productive places, samples of which are needed for truly representative conservation area networks (Sarkar 1999; Margules & Pressey 2000).

5.2 The goals of networks: representativeness, persistence and economy

As discussed above, conservation area selection has tended in the past to be either opportunistic, ad hoc from the point of view of biodiversity conservation, or a response to some perceived external threat. Many of the earliest national parks, for example, Yellowstone (USA) and Royal, near Sydney (Australia), were chosen for their outstanding physical features and natural beauty. Many conservation areas throughout the world continue to be chosen for similar reasons, or, for example, because they protect particular rare species or wilderness values. While the protection of beautiful landscapes, rare species and wilderness values may be socially desirable and legitimate goals, from the perspective of *biodiversity* conservation, which is actually quite a different goal, they are ad hoc. Rare species, of course, are key components of biodiversity and generally confer high complementarity value on the areas in which they occur. But a focus on rare species alone can divert scarce resources and effort away from other deserving areas. It is estimated, for example, that less than half

Table 5.1 *The representation of regional ecosystems (REs) in the reserve system of Queensland (Australia)*

	Bioregion	Regional area (ha)	Protected area[a] (ha)	No. of REs in bioregion	No. of REs in protected areas	% REs in protected areas	% REs in >1 protected area[b]
1	Northwest Highlands	6 950 000	369 100	41	27	66	22
2	Gulf Plains	21 377 000	525 300	83	25	30	10
3	Cape York Peninsula	11 548 000	1 594 100	211	177	84	54
4	Mitchell Grass Downs	22 787 000	238 900	53	21	40	15
5	Channel Country	24 594 000	1 634 700	56	44	79	52
6	Mulga Lands	19 097 000	464 900	66	47	71	39
7	Wet Tropics	1 850 000	310 400	105	71	68	41
8	Central Queensland Coast	1 151 000	142 300	37	33	89	59
9	Einasleigh Uplands	12 808 000	226 900	46	26	57	11
10	Desert Uplands	6 882 000	153 800	58	25	43	9
11	Brigalow Belt	35 158 000	730 400	163	110	67	39
12	Southeast Queensland	8 231 000	341 800	145	125	86	49
13	New England Tableland	341 000	26 500	21	14	67	19
	TOTAL	**172 774 000**	**6 759 100**	**1 085**	**745**	**69**	**39**

[a] From Sattler & Williams (1999). Reproduced with permission from the Environmental Protection Agency Queensland, Australia.
[b] Protected areas current to February 28 1998.

of the designated and proposed Tiger Reserves in India may still contain tigers. The resources going into maintaining the other half might better be spent elsewhere.

This focus on goals other than biodiversity conservation has led to a very uneven representation of biological diversity in conservation area networks: such networks suffer from a lack of *representativeness*: they do not protect all biodiversity surrogates adequately. Table 5.1 documents the current uneven representation of natural features in reserve networks in Queensland (Australia). Sixty-nine percent of all regional ecosystems (a term for vegetation communities) are represented in reserves, but that leaves 31 % unrepresented. Further, only 39 % are represented in more than one reserve. Leader-Williams *et al.* (1990) showed that the widely used opportunistic approach to conservation area selection has resulted in the uneven and inadequate representation of natural features in national parks in England and Wales, and in Zambia, where park location is biased towards areas with tsetse fly infestations. Pressey *et al.* (1996) found that the pattern of representation of environmental units in protected areas in temperate forest ecosystems of a region of eastern Australia was strongly biased towards steep slopes and infertile environments. Cantú *et al.* (2004) reported a similar bias towards high elevation and poor soils in an analysis of existing and proposed conservation areas in Mexico. These are just a few examples of a phenomenon that is widespread and commonplace throughout the world.

Representativeness alone is obviously insufficient. Conservation planning must also aim to ensure that the (true) surrogates for biodiversity now found in conservation areas persist into the future. The prognosis for many areas may be dim either because of anthropogenic threats (resource exploitation, development, etc.) or because of biological and environmental factors (competition from other species, climate change, etc.). Thus, *persistence* must be a central goal of conservation planning. We will return to the question of persistence in the next chapter.

Meanwhile, the area of land that will ever be set aside solely for biodiversity conservation is limited. Some years ago, a generally accepted target was 5 % of the total land area of a region. (This use of "target" must be kept distinct from "targeted" areas to be put under some conservation plan – this is a target for the amount of area or the level of representation of each biological feature; it does not refer to an individual area.) WWF and UQCN (1996) have argued for a minimum of 10 % of all forest types globally to be represented in conservation area networks. It seems that in many parts of the world it will be difficult to achieve even that, though by 1999, 20 countries had committed to this goal through the Convention on Biological Diversity and some, for example Canada (12 %; Hummel 1995) and Australia (15 %; Commonwealth of Australia 1997) had committed to exceeding it in at least some ecosystems (Kanowski *et al.* 1999) and much higher targets have occasionally been proposed (e.g., Ryti 1992). However, as discussed above, the adequacy of a conservation area network cannot be measured by the area of land it contains. Adequacy depends on the extent to which conservation area networks sample the range of natural variation in terms of species, communities, environments, etc. in a region, and on whether the biotic features are likely to persist into the future. For example, Colombia covers about 1 140 000 km^2. It has 49 national parks that together cover 100 983 km^2 or about 8.9 % of the

Table 5.2 *The representation of biodiversity in the reserve system of Colombia*

Topological representativeness[a]	Ecosystems declared protected	
	Number	%
0%	159	47.2
1–5%	61	18.1
5–10%	14	4.2
10–25%	35	10.4
25–50%	33	9.8
> 50%	35	10.4
TOTAL	**337**	**100.0**

[a] Topological representativeness = area protected/area of the original potential extent of the ecosystem expressed as a % (Fandiño-Lozano 1996).

country. That figure is approaching 10%. However, as Table 5.2 shows, Colombia's national parks do not form a representative protected-area network. Of 337 ecosystems recognized in Colombia (Fandiño-Lozano 1996) 159 are not represented at all, a further 61 have only between 1 and 5% of their original extent protected, and so on. Only 35 have more than 50% of their original extent protected.

Thus, it is necessary to design conservation-area networks that are *economical*; that is, new conservation areas must be selected carefully so that limited resources are deployed most efficiently (Pressey *et al.* 1993). (What is being called "economy" here has often been called "efficiency" in the literature. Efficiency implies that goals are achieved with least cost or effort. Economy is a better term because it allows the incorporation of multiple goals, the achievement of which may not be the most efficient and while efficiency is important, we might often be prepared to relinquish high efficiency for a more implementable solution as discussed in Chapter 6. In this book we reserve the term "efficiency" for the computational speed of algorithms.)

Representativeness can only be non-arbitrarily assessed provided that an explicit target of representation is set for each surrogate (either a certain number of times or a certain fraction of its occurrence in the data set). Representativeness can then be viewed as measured by the fraction of surrogates that meet their set targets. Our present biological knowledge does not entirely determine what targets are appropriate for each surrogate.

Rather, educated intuitions provide rules of thumb: for instance, low targets may be sufficient for common species under no apparent threat of extinction. For highly endangered species, the appropriate target may be every occurrence (Sarakinos *et al.* 2001). Soulé and Sanjayan (1998) have argued that targets such as 10% and 12% mentioned earlier are far too small to maintain viable populations of many species, in particular wide-ranging ones, and have only been adopted for political expediency. According to them, the achievement of such minimal targets may lead to an unjustifiable complacency about the prognosis for biodiversity. However, targets can be chosen to meet the exact needs of the biological entity of interest. Without explicit targets, the representativeness of a network cannot be gauged and complementarity cannot be implemented (Section 5.3 below). We will return to the problem of setting targets in the next chapter (Section 6.3).

5.3 Selecting networks: complementarity

A key concept in identifying areas to achieve representativeness economically is *complementarity*. This concept, which was already introduced in Chapter 1, underpins the examples described in detail in Chapter 8 though the term itself was not coined until fairly recently (Vane-Wright *et al.* 1991; Pressey *et al.* 1993). Conservation areas should complement one another in terms of the features they contain, the species, communities, habitats, etc. Each conservation area should be as different from the others as possible until all "differences," different species, communities, habitats, etc., are adequately represented. This sounds like plain common sense, especially with the benefit of hindsight, but as already mentioned above, conservation areas usually have not been chosen to complement one another in the past and this practice continues today.

Complementarity can be precisely defined in a variety of ways. The simplest one is that one area has a higher complementarity value than another if it has more surrogates that have not already met their assigned target of representation in a conservation area network. (If the target is one representation of all surrogates, then the complementarity value is the number of *new* surrogates an area would add.) Complementarity can be contrasted with richness, the number of surrogates that an individual area has. Two areas, both with high richness, can have very similar sets of surrogates. Including both in a conservation area network will not be economical if a third area with different surrogates is available and can be selected instead. The use of complementarity would preclude both

Example 5.2 Selecting wetlands in the Macleay Valley, New South Wales

One of the earliest algorithms for implementing complementarity was developed by A. O. Nicholls and used to select a subset of wetlands in the Macleay Valley, New South Wales. These wetlands were chosen to represent all plant species recorded in field surveys across all wetlands (Margules *et al.* 1988). There were four rules:

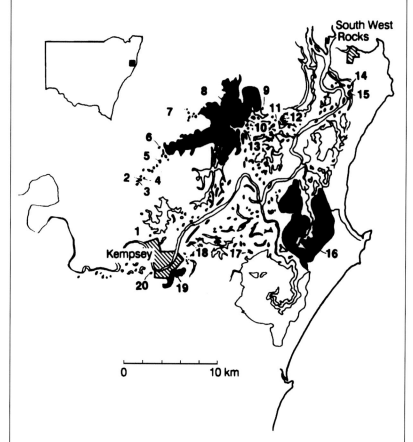

Figure 5.1. Wetlands of the Macleay Valley showing the 20 selected to represent all species at least once numbered and in black. The townships of Kempsey and South West Rocks are also shown. (From Margules *et al.* (1991). © CSIRO Australia 1991.)

(1) Select all wetlands with any species that occur only once.

(2) Starting with the rarest (i.e., least frequent species in the data matrix) unrepresented species, select from all wetlands on which it occurs, the wetland contributing the maximum number of additional unrepresented species.

(3) Where two or more wetlands contribute an equal number of additional species, select the wetland with the least frequent group of species. The least frequent group was defined as that group having the smallest sum of frequencies of occurrence in the remaining unselected wetlands.

(4) Where two or more wetlands contribute an equal number of infrequent species, select the first one in the list.

This final step is order dependent (and, thus, essentially random) and it was recognized that other criteria could be introduced here, as has been done many times subsequently. Figure 5.1 is a map showing all wetlands and those selected to represent all species at least once. Only 20 of the 432 wetlands were required to do this, but they accounted for almost 50 % of total wetland area. These numbers increased rapidly when more than one representation of each species was required reaching 65 wetlands totalling 78.5 % of wetland area for five representations of each species.

of the first two areas from inclusion in the presence of the third. Complementarity is thus related to what ecologists have called beta-diversity, the difference in feature content between two areas (Whittaker 1960). However, whereas beta-diversity is usually interpreted as the symmetric difference between two areas, complementarity is an asymmetric measure, describing how many new features an area adds to what is already available in a set of selected areas. Complementarity values, unlike traditional measures of beta-diversity, must be updated iteratively as a network of selected areas is constructed.

The introduction of the principle of complementarity (Kirkpatrick 1983; Ackery & Vane-Wright 1984; Margules & Nicholls 1987; Margules et al. 1988; Rebelo & Siegfried 1990) was one of the most important innovations in the design of conservation-area networks. By now complementarity has come to be used in most algorithmic procedures for conservation area network selection. Justus and Sarkar (2002) provide a history of its use until 2000 in contexts where, beyond the analysis of data sets for academic purposes, practical policy formulation was at stake.

Example 5.3   Ad hoc reserve-network selection in New South Wales

Pressey (1990) documented the consequences of opportunistic conservation-area selection in the north-western part of the Western Division of New South Wales (Australia). This is arid to semi-arid (rangeland) country used mainly for grazing sheep on unimproved native pasture. The land tenure is all leasehold. The areas in the data set were 1026 pastoral leases (ranches) and the biodiversity surrogates were 128 land systems: environmental classes representing recurring patterns of landforms, soils and vegetation (Christian & Stewart 1968). The New South Wales National Parks and Wildlife Service commenced a program of reserve acquisition in this part of the state in 1971. By 1988 reserves occupied 3.3% of the area. This accumulation is represented by the lower line in Figure 5.2. The upper line begins at 5.7%; this represents the total area required in 1971 if complementarity had been used to select a minimum set of pastoral leases to represent all land systems at least once. By 1988, and including the contribution of the reserves acquired since 1971, that figure had risen to 8.3%. The reserves acquired since 1971 contribute little to complete

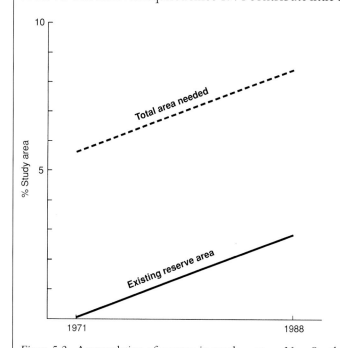

Figure 5.2. Accumulation of reserves in north-western New South Wales between 1971 and 1988 (lower line) and the total area needed in 1971 and 1988 to represent all land systems (environmental classes). See text for detailed discussion. (Re-drawn from Pressey (1990). Reproduced with permission from Zoological Society of NSW, Australia.)

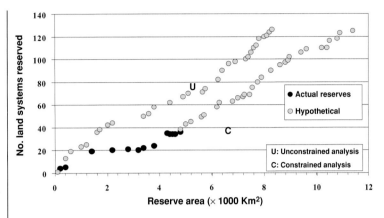

Figure 5.3. Accumulation of land systems (environmental classes) in actual reserves and in hypothetical reserves chosen using complementarity, in north-western New South Wales. See text for detailed discussion. (Re-drawn from Pressey (1990). Reproduced with permission from the Zoological Society of NSW, Australia.)

representation because they have not added many new land systems: the two lines in Figure 5.2 are almost parallel.

Using the same data set, Figure 5.3 shows the rate of accumulation of land systems in a hypothetical conservation area network selected using complementarity, which ignores existing reserves (unconstrained analysis), and in another possible reserve network, which builds on existing reserves (constrained analysis). If existing reserves are ignored, all 128 land systems can be represented in 7980 km^2. Starting with existing reserves, the accumulation curve is shifted laterally and the minimum area then required to represent all land systems, using complementarity to identify new reserves, is 11 503 km^2. If future conservation areas are not chosen explicitly to complement existing ones, this curve will flatten even further and the acceptable limit of total reserve area (whatever it is) may be reached before all land systems are represented in the reserve network.

Pressey (1990) acknowledged that in practice neither of the curves shown in Figure 5.3 is achievable. Existing reserves cannot be ignored and the pastoral leases representing the best accumulation on top of existing reserves needed to achieve full representation are unlikely to become available when required. His point was simply that complementarity should inform decisions on reserve acquisition so that maximum representation can be obtained wherever possible.

Table 5.3 *The number of 10 km² grid cells and the numbers of breeding birds selected using richness, rarity and complementarity, using data from the Atlas of Breeding Birds of the British Isles*[a]

	Area-selection methods			
			Complementarity areas	
Representation achieved (%)	Richness hotspots	Rarity hotspots	one representation	six representations
Occupied grid cells chosen ($n = 2887$)	5.0 (141)	5.0 (141)	1.0 (27)	4.9 (139)
Total species represented ($n = 218$)	89.0 (194)	97.7 (213)	100.0 (218)	100.0 (218)
Total number of records ($n = 170\ 098$)	7.8 (13 208)	6.1 (10 329)	1.1 (1954)	6.0 (10 141)

[a] From Williams *et al.* (1996). Reproduced with permission from Blackwell Publishing.

Example 5.4   Breeding birds in the British Isles

Williams *et al.* (1996) analyzed breeding bird data from 10 km² grid cells throughout the British Isles. They compared the results obtained by selecting the 5% of richest cells, the 5% of cells with the rarest species and sets chosen using complementarity. Table 5.3 shows that if the 5% of richest cells are chosen, which is 141 grid cells, then 89% of species are represented (194 species). If the 5% of cells with the rarest species are chosen, then nearly 98% of species are represented (213 species). However, if complementarity is used, then all 218 species can be represented at least once in just 1% of grid cells consisting of 27 cells. If it is possible to afford 5%, then all species can be represented six times in 4.9% of grid cells consisting of 139 cells. Of course it is only possible to represent species six times if there are at least six breeding occurrences. If there is only one, then that one is represented, if two then both are represented and so on up to six. Clearly, using complementarity to select a set of grid cells to represent all breeding birds is far more cost-effective than using richness or rarity.

Example 5.5   The protected area network of Québec

Sarakinos *et al.* (2001) developed a nominal conservation-area network for Québec. In 1999 only 4.2% of the land area of Québec was under some form of protection, with the provincial government committed to a goal of putting 12% of the area under a conservation plan by 2000 (Hummel 1995). The study area (about $1\,522\,842\,\mathrm{km}^2$) was divided into 21 403 cells at a $0.2° \times 0.2°$ longitude $\times$ latitude resolution. As (true) surrogates for biodiversity they used 400 faunal and floral species at risk (346 plant species and 54 animal species), 22 native small mammal species, 6 game mammal species and 92 fish species. The data were a mixture of presence-only and presence–absence records.

Figure 5.4 shows the distribution of species at risk in Québec. It shows that, except in southern Québec, species at risk are represented poorly in the existing protected areas. Only 118 of the plant species at risk (or 34.1%) and 20 of the animal species at risk (or 37.0%) had records within the protected areas. This result should come as no

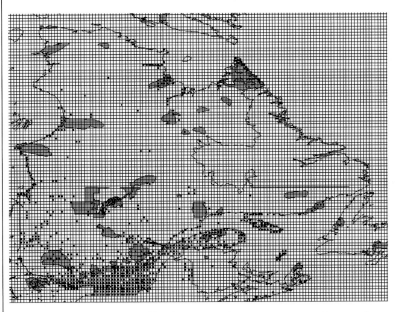

Figure 5.4. Representation of biodiversity in the existing protected areas of Québec. Distribution of species at risk in Québec (black dots). The grey areas are the existing protected areas. Most records of species at risk fall outside the grey areas. (From Sarakinos *et al.* (2001)). Reproduced with permission from Kluwer Academic Publishers (Springer Science).)

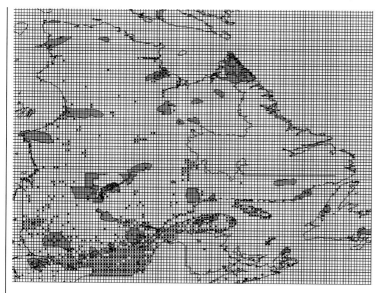

Figure 5.5. Area selection using complementarity in Québec. Distribution of selected areas in Québec (black dots). The targets were 50 representations for species at risk and 100 representations for all other species (small and game mammals, and fish). The grey areas are the existing protected areas. Note how almost all the selected areas fall outside the grey areas. (Redrawn using data from Sarakinos *et al.* (2001). Reproduced with permission from Kluwer Academic Publishers (Springer Science).)

surprise since most of the protected areas were known to be selected for their scenic or recreational (especially skiing) value.

Figure 5.5 shows the areas selected using a complementarity-based algorithm (and breaking ties by rarity and adjacency). The algorithm for selecting cells had up to seven steps:

Rule 0 – (Optional initialization) Select cells that are in an existing protected-area network;

Rule 1 – (Complementarity) Select a cell or cells which contain the most surrogates with an individual frequency less than or equal to the targeted level of representation;

Rule 2 – (Rarity I) If a choice exists, select a cell or cells which adds the next rarest under-represented surrogate and maximizes the additional representation of currently under-represented surrogates;

Rule 3 – (Optional adjacency rule) If a choice exists, select a cell or cells which is adjacent to one of the cells that have already been selected;

Rule 4 – (Rarity II) If a choice exists, select a cell or cells which adds the next rarest under-represented surrogate and contains the most infrequent additional under-represented surrogates;

Rule 5 – (Rarity III) If a choice exists, select a cell or cells which adds the next rarest under-represented surrogate and contains the most infrequent additional surrogates;

Rule 6 – (Random/ lexical order rule) If a choice exists, select the first cell in the list which adds under-represented surrogates.

In Figure 5.5, a target of 50 representations was set for each of the species at risk, and 100 representations for the other species. Rules 0 and 3 were not used. Sarakinos *et al.* found that the representation of species at risk within the existing protected areas was so poor that results of selection initiated with the existing protected areas and selection *ab initio* did not lead to significantly different new cells being selected.

This example will be further discussed in Chapter 8 (Section 8.2).

5.4   Selecting networks: rarity and adjacency

When targets are only set in terms of the representation of each attribute, complementarity alone suffices to provide an economical solution to the problem of representativeness. However, on occasion, an additional target or constraint is set, such as the total land area that can be put under some conservation plan (see Section 5.3). Complementarity does not explicitly distinguish between rare and common attributes. Therefore, when a selection procedure is terminated because the area target is met, if complementarity alone has been used, it is possible that rare attributes (which are often of greater conservation value because of their higher risk of extinction) have been excluded in favor of common ones, especially if the target falls short of representing all species. Many procedures for conservation area network selection attempt to guard against this eventuality by using rarity along with complementarity, sometimes giving the former precedence (Sarkar *et al.* 2002).

The rarity of a surrogate can be defined using three criteria, either individually, or in combination with each other: (1) geographical range – rarity is the inverse of the frequency with which a surrogate occurs in a set of places, that is, the inverse of the size of its geographical range; (2) habitat specificity – the more specific a surrogate's habitat is, the rarer it is; and (3) local population size – the rarity of a surrogate is the inverse of its local abundance (Rabinowitz *et al.* 1986). There can thus be seven different types of rarity. These criteria are not independent of each other: for instance, if habitats are limited, rarity defined by geographical range will tend to coincide with rarity defined by habitat specificity.

So far, only rarity as defined by geographical range has been routinely used in the design of conservation area networks. The most important reason to use rarity as defined by geographical range, especially along with the use of complementarity, is that it enables a preference for endemic species to be incorporated into selection procedures when area targets are used. Endemism is a crucial criterion in the definition of "hotspots" of biodiversity (Myers 1988, 1990). Thus, using rarity encourages the inclusion of hotspots in conservation area networks. Csuti *et al.* (1997) found that, for binary data (that is, with one indicating presence and zero indicating absence), algorithms combining rarity and complementarity typically performed best with respect to economy (see, also, Sarkar *et al.* 2002). However, when data are probabilistic, complementarity alone produces the most economical plans (Sarkar *et al.* 2004b).

Complementarity and rarity are the most important rules used to select conservation areas for inclusion in networks. Another rule that is often used is *adjacency*: all else being equal, preference is given to an area adjacent to those already in a (potential) conservation area network over those that are not (see Figure 5.4). The effect is to have larger individual conservation areas, which may be important if the planning units used are small and, therefore, likely to have populations of species that may not persist in the long run. Clearly it does not make sense to give this rule higher priority than complementarity and rarity.

5.5   Subsidiary goals: flexibility, transparency, modularity, genericity and irreplaceability

The selection of conservation area networks is not a one-off process. Proposed networks must be accepted and implemented by planning bodies, which bring many economic, political and social considerations to bear upon decisions. (Chapter 7 will discuss methods for the incorporation of some of these considerations in the refinement of conservation-area networks.) It is, therefore, critical that there should be several alternatives available which are equally adequate from the point of view of the economical representation of biodiversity surrogates in them. This is the concept of *flexibility* (Pressey *et al.* 1993; Church *et al.* 1996).

Flexibility requires, first and foremost, that replacements be available for individual areas that are precluded from inclusion in a conservation area network because of considerations such as those mentioned in the last paragraph or for some other reason, for instance, a poor prognosis for the persistence of biota in that area. Thus, it must be clear why each individual area is selected, for instance, whether it was selected because of the presence of some rare attribute or because of complementarity. This is known as *transparency* (Nicholls & Margules 1993). Transparency is important because, should such an area be removed from a network, planners must be able to assess exactly what features are lost so that they can be regained elsewhere, for example, through the selection of an additional area.

Moreover, flexibility is enhanced if each rule that is used in the design of the conservation area network is independent of the others, so that some can be selectively ignored in some contexts. This is known as *modularity* (Sarkar *et al.* 2004b). For instance, it should be possible to use adjacency selectively, only when the conservation areas are small or when there is reason to believe that connectivity between conservation areas in a network is important. Finally a conservation area selection procedure should be able to resolve data sets from anywhere in the world. It should not be specific to some attribute (for instance, taxon) or region. This is known as *genericity*.

Economic, social or cultural considerations often preclude the implementation of a complete conservation area network as designed. Rather, the usual response to a design is to formulate a sequential stage-by-stage plan of implementation. However, this strategy leaves open the possibility that areas to be included at later stages become transformed into biological irrelevance during the waiting period. Consequently, it is important to identify those areas that can be more readily substituted by other areas and those for which there are few, or no, substitutable areas. Though substitutability is probably the more appropriate term, *irreplaceability* has come to be used in its stead (Pressey *et al.* 1994). A high degree of irreplaceability corresponds to a low degree of substitutability.

Faced with a sequential implementation strategy, the recommendation of planners to decision-makers may be to designate areas for conservation action in order of their irreplaceability. However, two important caveats must be kept in mind regarding the use of irreplaceability:

Example 5.5   A conservation plan for Guyana

Guyana, with an area of $215\,000\,km^2$, is unique as the only country in the Americas without a system of protected areas. It includes biologically diverse environments, such as white-sand forests, savannah and Amazonian rainforests (Richardson & Funk 1999). Facing international pressure to harvest its forests and minerals in an environmentally sensitive fashion, the Government of Guyana ratified a National Environmental Action Plan (NEAP) in 1994 and agreed to begin the process of designating a network of conservation areas. A database with 16 500 records of 312 species covering ten taxonomic plant and animal groups was created with the help of the Smithsonian Institution. The spatial distribution patterns of all species with more than ten occurrences were modeled using DOMAIN (Carpenter *et al.* 1993).

These data were mapped to 941 $16 \times 16\,km^2$ cells and targets of representation were set to 15 % of the predicted distribution area for each species. Sites were selected using complementarity, and 140 sites were required to represent at least 15 % of the modeled ranges of all species. Of these, 36 had an irreplaceability value of one (K. S. Richardson, personal communication). However, plans to develop a conservation area have stalled even though the Global Environment Facility (GEF), funded through the World Bank, and other financiers were willing to provide $9 million for the initial implementation of two conservation areas (Justus & Sarkar 2002). The planners had not taken into account the land tenure practices and rights of Amerindian communities living in the designated areas. These communities had not been consulted. Consequently, it proved impossible to develop a politically feasible plan for implementation.

(1) The irreplaceability of an area can only be defined relative to a set of solutions. For instance, given a set of equally economical solutions, suppose that an area appears to have high irreplaceability because it is the only one that could add some feature to a conservation area network. It may still be the case that there are areas which do not occur in the solution set that also could add that feature. Thus, the irreplaceability rank of an area must be relativized to the set being considered, in this case, the solution set;

(2) The irreplaceability rank of an area will change as individual areas are designated as part of the conservation area network. Consequently, the process of selecting potential areas for the network must be re-iterated after each stage, when some new areas are included in a network. The complementarity value of an area must also similarly be iteratively updated (see Section 5.4). However, while the complementarity value is easy to compute, the irreplaceability value of an area can very be difficult to compute in practice (Tsuji & Tsubaki 2004).

An extended discussion of the concepts introduced in Sections 5.4 and 5.5 can be found in Sarkar *et al.* (2004b).

5.6 Algorithms for the selection of networks

As just noted, all the methods for selecting conservation-area networks discussed earlier in this chapter use heuristic rules with transparent biological interpretations. By design, these methods are computationally efficient and can resolve large data sets in reasonable amounts of time. After Margules *et al.* (1988) produced the first computer implementation of a complementarity-based algorithm, Cocks and Baird (1989) pointed out that the optimization technique of mathematical programming can be used to find exact or "optimal" solutions to network selection problems.

Formally, there are two related problems to be solved which can be interpreted as the following: given a list of cells, $\Sigma(\sigma_j \in \Sigma, j = 1, 2, \dots, n)$ representing the region; a list consisting of the areas $\alpha_j (j = 1, 2, \dots, n)$ of each of the cells; a list of (estimator) surrogates, $\lambda_i (\lambda_i \in \Lambda, i = 1, 2, \dots, m)$ for biodiversity; a target, $\tau_i (i = 1, 2, \dots, m)$, for each surrogate; and an expectation, $p_{ij} (i = 1, 2, \dots, m; j = 1, 2, \dots, n)$, of finding λ_i (the i-th surrogate) at σ_j (the j-th cell), we can solve:

(1) **Area minimization problem**: select the set of cells, Γ, with the minimum total area such that every surrogate meets its assigned target;
(2) **Representation maximization problem**: given a maximum total area, A, for the set, Γ, of selected cells, maximize the expected number of surrogates that meet their targets within A.

In the area minimization problem, economy is at stake but there is no constraint on the total number of cells (up to n) that can be selected. In the representation maximization problem, a constraint is present in the form of an upper limit to the area of the cells that can be conserved.

Let the variables $X_j (j = 1, 2, \ldots, n)$ and $Y_i (i = 1, 2, \ldots, m)$ be defined as follows:

$$X_j = \begin{cases} 1, & \text{if cell } \sigma_j \in \Gamma \\ 0, & \text{if cell } \sigma_j \notin \Gamma \end{cases} \quad \text{and} \quad Y_i = \begin{cases} 1, & \text{if } \sum_{\sigma_j \in \Gamma} p_{ij} > \tau_i \\ 0 & \text{otherwise} \end{cases}$$

X_j identifies (by the value 1) the selected cells while Y_i identifies those surrogates that have achieved their targets.

The area minimization problem then consists of the problem:

$$\text{Minimize } \sum_{j=1}^{n} \alpha_j X_j \text{ such that } \sum_{j=1}^{n} X_j p_{ij} \geq \tau_{ij} \text{ for } \forall \lambda_i \in \Lambda$$

The sum being minimized is the total area of the selected cells (noting that X_j is only non-zero if the j-th cell is selected). The constraint ensures that all targets are met within the selected cells. Note that any cost associated with each cell can be used instead of the area and similarly minimized.

The representation maximization problem then consists of the problem:

$$\text{Maximize } \sum_{i=1}^{m} Y_i \text{ such that } \sum_{j=1}^{n} \alpha_j X_j \leq A$$

The sum being maximized is the number of surrogates that have achieved their targets. The constraint is a budget limiting how many areas can be acquired, a problem almost always faced when conservation plans have to be implemented. (Again, any cost and not only area, can be incorporated into this description of the problem.)

The goals of conservation planning explicitly address the future status of biodiversity. Consequently, ideal data sets would consist of the probabilities of the persistence of biodiversity surrogates within areas at some future projected time. For instance, the future probabilistic expectation of the presence of a species could be its probability of persistence at a specified time in a given area. Alternatively, it could be its probability of colonization of a given area by that time. However, in the conceptual discussions of this book we will usually assume that what are available are binary data (with one signifying the presence of a biodiversity attribute and zero its absence) and that the data are presence–absence. Extending these concepts and methods to probabilistic data is an active ongoing program of research with many unresolved issues (Camm *et al.* 2002; Sarkar *et al.* 2004b). The problem of using presence-only data instead of presence–absence data has already been discussed in Chapters 3 and 4.

Both the area minimization and the representation maximization problems have optimal algorithms for their solution, which have been extensively studied by the operations research community. As far as economy is concerned, these optimal algorithms necessarily perform better than heuristic algorithms. However, these algorithms suffer from poor computational efficiency, that is, they may take inordinate amounts of time to resolve realistically sized data sets. Consequently, the gain in economy may not offset the cost in efficiency in most practical contexts, besides not being transparent (Pressey *et al.* 1996). Extensive numerical tests in the 1990s supported the last claim and underscored that, compared to heuristic rules, especially rarity and complementarity, the optimal algorithms only achieve a minor increase in economy at the cost of a significant loss of efficiency in the form of increased running time (Csuti *et al.* 1997; Pressey *et al.* 1997).

Rodrigues and Gaston (2002) and Rodrigues *et al.* (2000) questioned these results and claimed that new software solvers for mathematical-programming problems produced optimal solutions as efficiently as heuristic algorithms. However, Sarkar *et al.* (2004b) examined these claims systematically. Using the industry-standard integer programming CPLEX and other solvers, they obtained even more computationally efficient optimal solutions than Rodrigues and Gaston (2002). However, they found that heuristic algorithms based on complementarity outperformed optimal solvers with respect to efficiency with little loss of economy for realistically sized data sets with field (as opposed to artificial) data, especially if the data are probabilistic rather than presence–absence.

Besides heuristic and optimal algorithms, there exists a third class of algorithms that can potentially be used for the selection of conservation-area networks. These are called meta-heuristic algorithms and, when successfully implemented, achieve almost as much economy as optimal algorithms without sacrificing much of the efficiency of heuristic algorithms. A heuristic algorithm is a one-pass algorithm carrying out a sequence of steps once (see Examples 5.2 and 5.5). A meta-heuristic algorithm controls the execution of such a one-pass heuristic applying it several times until some other independent exit criterion is satisfied. For instance, we could apply complementarity to select areas until we satisfy all representation targets, then drop a subset of areas, and apply complementarity again to see if we do any better. The independent exit criterion could be the total permitted time, the number of "passes" through the heuristic, etc. Well-known meta-heuristic algorithms include genetic algorithms, neural networks, scatter search, simulated annealing and tabu search. So far, only simulated annealing (Kirkpatrick *et al.* 1983) has been frequently used for the selection of conservation area networks (Ball & Possingham 2000). An example, including the way in which simulated annealing was implemented, is discussed in Chapter 8 (Section 8.3).

5.7 The trouble with scoring and ranking procedures

Alternative methods of assigning conservation value usually involve a process of scoring areas against a set of criteria and then ranking them according to an accumulation of scores. The most common examples of criteria that have been used for scoring and ranking are diversity, rarity and representativeness, although there are many others, reflecting concepts such as ecological viability, the likelihood of external threat, naturalness, size, connectedness, etc. (Margules & Usher 1981; Smith & Theberge 1986; Usher 1986; Chomitz *et al.* 1999; Environmental Protection Agency, 2002; Oliver & Parkes 2003; Parkes *et al.* 2003). These criteria all reflect defensible scientific, esthetic or ethical concerns, but the application of them is not economical and may well perpetuate the uneven representation of natural features in conservation-area networks described in Section 5.2 above.

The question is, how far down the list is it necessary to go before the conservation-area network is deemed to be adequate? Pressey and Nicholls (1989) asked this question of the same data set that was used in Example 5.2 above. They identified eight separate criteria of conservation value encompassing aspects of rarity, representativeness, diversity and size, and calculated a score for each pastoral lease, according to each criterion. They also included four combinations of criteria for a total of 12 different scores for each lease. They then ranked the pastoral leases from the highest score to the lowest on each criterion or combination of criteria and asked how far down each list they would need to go before all land systems were represented once, twice, three times, four times and five times. Figure 5.6, re-drawn from Pressey and Nicholls (1989), shows the results. It also includes the results of two different selection methods using complementarity.

The relative economy of the different scores varied widely, but they were all considerably less economical than the complementarity based procedures. Even the most efficient of the scoring and ranking procedures required over 45 % of the total area of the region for at least one

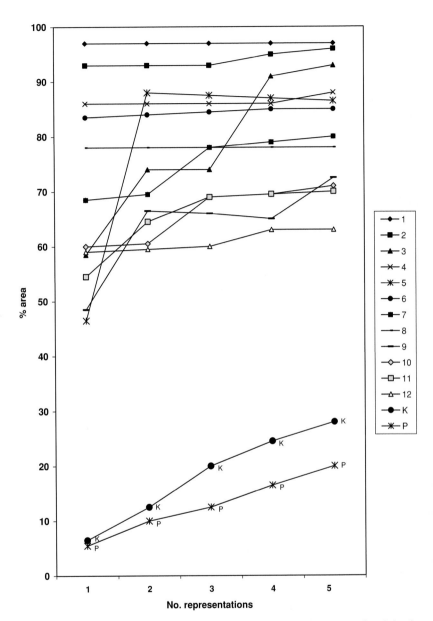

Figure 5.6. Percent area required to achieve up to five representations of each land system in sets of pastoral leases in north-western New South Wales. K and P are complementarity based analyses of Kirkpatrick & Harwood (1983) and Pressey and Nicholls (1989) respectively. The other 12 lines represent the areas required to represent each land system from one to five times if scores from criteria or combinations of criteria of conservation value are used to rank pastoral leases from highest to lowest. See Pressey & Nicholls (1989) for details of the criteria used for this analysis. (Re-drawn from Pressey & Nicholls (1989). Reproduced with permission from Elsevier.)

representation of each land system. It is obviously uneconomical to seek 45% or more in order merely to represent each land system at least once, when a complementarity-based algorithm can identify 20% of the area that contains at least five representations of each land system. If representation is not a goal, maybe it doesn't matter. But if representation is a goal, then it matters a great deal.

The question of the size of individual conservation areas is not the issue here; rather, the total area of the entire conservation area network. Individual components of the network, the conservation areas themselves, may be large or small and require different levels of management accordingly. The Western Division of New South Wales occupies approximately 320 000 km² and pastoral leases vary enormously in size. It may be that a smaller number of large reserves, insufficient to represent all features initially, will stand a higher probability of retaining those features over time. A decision along these lines would require the setting of priorities for different species and a reasonably confident prediction of the viability of populations of those species. Unfortunately, the only valid test of any such predictions is to wait and see. These issues are discussed in the context of persistence and vulnerability in Chapter 6. At this stage, the purpose is to establish the underlying rationale for complementarity-based procedures, which can and should, where possible, incorporate probability of persistence along with a range of other constraints.

5.8   Summary

The role of conservation area networks is to represent biodiversity (i.e., the chosen biodiversity surrogates) and ensure as far as possible that biodiversity persists within that network of areas. Two key goals of networks are therefore representativeness and persistence. A third key goal is economy. Economy means that new conservation areas should be chosen in a way that deploys limited resources most efficiently. When this has not been done in the past, conservation area networks have not represented biodiversity well. Complementarity is used to achieve representativeness economically. An area has higher complementarity than another if it has more biodiversity surrogates that have not already met their targets in conservation areas previously selected. Rarity is often used with complementarity to ensure that all rare surrogates are adequately represented. In this case, the selection of areas begins with the area with the largest number of rare surrogates and proceeds to select areas with the rarest unrepresented surrogates, and so on until all surrogates are represented.

Adjacency can be used in a similar way. If two areas have the same complementarity value, the one closest to an area previously selected can be given priority to ensure that distances between conservation areas are minimized.

There are a further five subsidiary goals: *flexibility*, substitutes are available where possible; *transparency*, area choices should be for explicit reasons; *modularity*, rules in selection algorithms are independent; *genericity*, area selection procedures should be applicable anywhere in the world; and finally *irreplaceability*, which refers to the extent to which areas can be substituted for others. A high degree of irreplaceability means a low degree of substitutability.

Heuristic algorithms are now widely used to select conservation priority areas. These methods are computationally efficient and can resolve large data sets in reasonable amounts of time. However, there are optimization techniques such as mathematical programming that can also be used to implement complementarity, but they are less efficient. Alternative methods of assessing conservation value by scoring each area using a set of criteria such as richness, diversity and rarity, or combinations of these, are grossly uneconomical.

Plate 1. 608 environmental domains in Papua New Guinea. Similar colours represent similar domains. The data described in Chapter 3 (Example 3.1) were used to derive these domains, as described in Chapter 4 (Example 4.1). The use of the domains in conservation planning is described in Chapter 8 (Section 8.5). (Reproduced with permission from the Centre for Resource and Environmental Studies, Australian National University.)

Plate 2. Wet sclerophyll forest in south-eastern New South Wales with *Eucalyptus fastigata* trees, and tree ferns, *Dicksonia antarctica*, in the understory (Example 7.3). (Photo Liz Poon.)

Plate 3. Milne Bay Canoe Festival 2004, Alotau, Papua New Guinea. (Examples 3.1 & 4.1 & Section 8.5). (Photo courtesy of Conservation International.)

Plate 4. Forest clearing for gardens. Diodio Village, south-west Goodenough Island (Moratua), Papua New Guinea. (Examples 3.1 & 4.1 & Section 8.5). (Photo David Mitchell, Conservation International.)

Plate 5. Spinifex, *Triodia sp.*, on a dune crest on Cuthero Station, south of Broken Hill in the Western Division of New South Wales, Australia (Sections 2.2.1, 5.3.2 & 5.7). (Photo Liz Poon.)

Plate 6. Black Box, *Eucalyptus largiflorens*, along the Darling River, the major drainage in the sparsely watered Western Division of New South Wales (Sections 2.2.1, 5.3.2 & 5.7). (Photo Liz Poon.)

Plate 7. Rainforest canopy seen from a cable-car servicing telecommunications infrastructure on Mt. Bellenden Ker in the Wet Tropics of north-eastern Queensland. Species include *Elaeocarpus ferruginiflorus*, with the flush of new red growth, *Orites excelsa* and *Caldcluvia australiensis* (Examples 2.3 and 3.4). (Photo CSIRO.)

Plate 8. Hinchinbrook Island seen from the mainland with mangroves in the middle distance. Wet Tropics of north-eastern Queensland (Examples 2.3 and 3.4). (Photo Liz Poon.)

Plate 9. Zebra and Gemsbok at a waterhole in the Etosha National Park, north central Namibia (Examples 5.1 & 7.5). (Photo Chris Margules.)

Plate 10. Grain storage bins, north central Namibia (Examples 5.1 & 7.5). (Photo Chris Margules.)

Plate 11. Coastal wet forest along the coastal cordillera of Ecuador west of the Andes. This habitat type is currently poorly protected (Example 7.4.2). (Photo Rodrigo Sierra.)

Plate 12. Mangrove swamps at sea level in eastern Ecuador, threatened by shrimp farming. This is a priority area identified by Sarkar *et al.* (2004a) (Example 7.4). (Photo Rodrigo Sierra.)

Plate 13. Jenn Caselle, Partnership for Interdisciplinary Studies of Coastal Oceans (PISCO)/ University of California at Santa Barbara (UCSB) science coordinator at Anacapa Island, Channel Islands of California (Section 8.3). (Photo Brad Doane.)

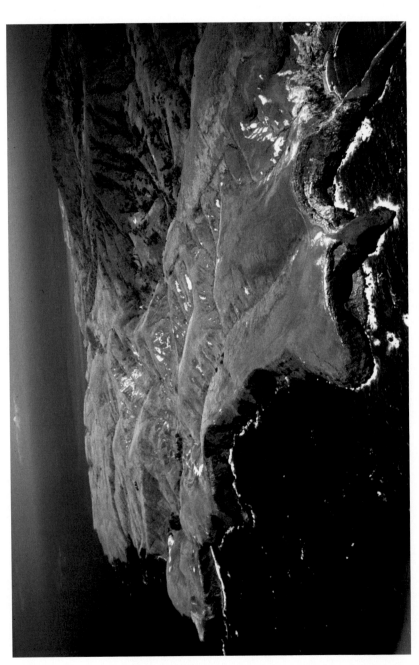

Plate 14. Aerial view of the Channel Islands of California (Section 8.3). (Photo Peter Taylor.)

Plate 15. A sink hole in the calcareous crust of the Nullarbor Plain. Nullarbor is from the Latin for no trees. This plain slopes gently from north to south on a gradient of 1:200. Bluebush, *Maireana spp.*, are common dominant plants (Examples 3.3 and 4.2 and Section 8.1). (Photo Chris Margules.)

Plate 16. Near quadrat HA1 on the central Nullarbor Plain. This low shrubland is an example of species assemblage 9 (Examples 3.3 and 4.2 and Section 8.1). (Photo Norm McKenzie.)

Plate 17. Fynbos vegetation on nutrient-poor sands in the Table Mountain National Park on South Africa's Cape Peninsula. This peninsula, comprising 471 km², is home to 2256 plant species. Note the prevalence of Restionaceae in the foreground. The large shrubs are all members of the Proteaceae (Leucadendron, Protea) (Example 6.3 & Section 8.4). (Photo Richard Cowling.)

Plate 18. Fynbos vegetation on the Groot Winterhoek Mountains near the eastern boundary of the Cape Floristic Region. Here, owing to a higher incidence of summer rainfall than further west, fynbos is enriched by species from adjacent biomes, especially grasses from the subtropical grassland biome. The tall shrub the foreground is *Protea nitida* (Example 6.3 & Section 8.4). (Photo Richard Cowling.)

Plate 19. Landscape in the vicinity of the Ekati and Diavik diamond mines (Lac De Gras, Northwest Territories, Canada) in September. This area is traversed by caribou of the Bathurst herd making their southerly migration from the Arctic coast near Bathurst Inlet to the wintering range in the vicinity of Great Slave Lake, NWT (Example 6.2). (Photo Chris Johnson.)

Plate 20. Dry creek bed in Lakefield National Park, Cape York Peninsula, Australia. *Coix gasteenii* occurs along some of the seasonal drainage lines that feed into this creek bed (Example 4.3). (Photo Liz Poon.)

6 · Persistence and vulnerability

We noted in Chapter 5 that the central objectives of systematic conservation planning are representativeness, persistence and economy. So far we have focused on the selection of areas to achieve representativeness and economy in the design of conservation area networks. We also noted in Chapter 1 that identifying such areas will not by itself protect biodiversity. Identification is the first step in planning for the persistence of biodiversity (or, in the terminology of Chapter 2, the true surrogates for biodiversity). Planning for persistence requires attention to be paid to vulnerability; the potential for biodiversity features to be lost (Nicholls 1998; Wilson *et al.* 2005). For discussions of the general concept of vulnerability, see Cutter (1996) and Dilley and Boudreau (2001). In practice, conservation planners often concentrate on the potential for imminent loss because this is believed to require immediate action (Pressey *et al.* 1996; Conservation International 2004). However, it is equally plausible to hold the view that the potential long-term loss deserves at least equal attention (Caughley 1994).

The objective of persistence underlies all stages of systematic conservation planning (Table 1.1) though its influence on Stages 2 and 6 is relatively limited. Stakeholders whose actions have an impact on biodiversity persistence must be brought into the planning process in Stage 1. In Stage 2, data collection, if constrained by limited resources (time and money), should focus on biodiversity features with uncertain likelihoods of persistence. In Stage 3, species at risk are prime candidates for inclusion in true-surrogate sets. Persistence assumptions are central to setting goals and targets in Stage 4. The review of existing conservation areas in Stage 5 should include an assessment of the likelihood of the persistence of the relevant biodiversity features of these areas. In Stage 6, the selection of new conservation areas typically excludes those with very low likelihoods of persistence, for instance, anthropogenically transformed areas. Assessing these likelihoods more precisely is the task of Stage 7. In Stage 8, these assessments are then used to refine the results of Stage 6. As we will see later, correlates of persistence and vulnerability are among the measures used in the multi-criteria analysis of Stage 9. The implementation of conservation plans in Stage 10 should have persistence as a major explicit objective. Similarly, the reviews envisioned in Stage 11 should assess how successful the persistence of biodiversity has been in existing conservation area networks.

6.1 Incorporating biological processes

Planning for persistence requires attention to biological and socio-political factors beyond the simple representation of surrogates in conservation area networks. Persistence is dependent on both spatial and temporal scales. With respect to spatial scale, we must try to ensure that biodiversity features persist over the entire landscape. This means that we have to take into account ecological and other processes both inside and outside conservation area networks. For most biodiversity features, ideally, persistence should be ensured within the scale of individual conservation areas. However, in many instances, this may not be an attainable goal and attention must be paid to how processes outside such areas impinge on the areas themselves. Turning to temporal scale, factors threatening persistence may change through time in extent, rate, intensity and type (Gaston *et al.* 2002).

Taking all such factors into account over long timescales is usually impossible. Typically what we can do is try to decrease the potential for imminent loss over shorter timescales (Pressey *et al.* 1996). What the extent of this short-term planning timescale should be remains poorly understood. Several recent studies have assumed it to be about 50 years or slightly longer (Burgman *et al.* 2001), though population viability analysis sometimes assumes much longer timescales (see Section 6.2 below). It should be emphasised that the use of a short timescale does not presume that longer timescales are not important for biological processes, particularly evolutionary processes. Rather, it acknowledges large uncertainties in planning for persistence, and because such uncertainties tend to get amplified over time, quantitative planning at longer timescales is unrealistic. Nicholls (1998) explains how the kinds of methods described in Chapter 5 could be modified to incorporate information on persistence, if it is available.

Conservation planning has drawn on seven sets of ideas intended to safeguard the persistence of biodiversity surrogates after a conservation area network is established (Margules & Pressey 2000):

(1) Biogeographical theory – the theory of island biogeography (MacArthur & Wilson 1967) was extensively used to design nominal conservation areas in the 1970s (Diamond & May 1976; Harris 1984). Conservation areas were supposed to be analogous to oceanic islands. More species are predicted to occur on islands that are large and closer to the mainland than islands that are small and far from the mainland. Drawing on this theory and other ideas about edge effects and dispersal, the suggestion was that conservation area networks should consist of large circular reserves that are close together or close to larger continuous tracts of habitat ("mainlands") and be linked by corridors (e.g., Diamond 1975). Such a design is supposed to maximize the number of species that can persist in the network. However, caution must be exercised in applying these ideas to conservation area networks – there is no unequivocal evidence supporting the analogy between islands in oceans and terrestrial reserves (Higgs 1981; Margules *et al.* 1982) and the evidence that species use corridors for dispersal is also equivocal. In addition, the theory of island biogeography deals with size and distance, but nothing else. An early criticism, widely ignored by conservation biologists, was that the theory deals with characterless species on featureless plains (Sauer 1969). Attention has rightly shifted to the roles of environmental heterogeneity, species interactions, local- and regional-scale population dynamics and the effects of habitat modification on species persistence;

(2) Metapopulation dynamics – many species are distributed across landscapes as metapopulations (Hanski 1998; Cabeza & Moilanen 2001; Moilanen & Cabeza 2002; Bowne & Bowers 2004). Metapopulation theory calls for designs that aim for representation across species' ranges so that some populations might escape stochastic events such as wildfire, thus spreading the risk of extinction. This also suggests that, to ensure the persistence of such metapopulations, prioritization should include areas that establish connectivity between local populations to facilitate migration and dispersal. Corridors of habitat are one form of connectivity that might be useful for some species;

(3) Successional pathways – a conservation area network should represent different successional stages corresponding to (true) surrogates' habitat requirements (Pickett & Thompson 1978; Bengtsson *et al.* 2003). Much of conservation planning in the past has been preoccupied with communities that are characteristic of the end of successional sequences (Usher 1993) but the conservation of other stages is

equally important. Large conservation areas are better at meeting this objective since they are less likely to be entirely reset to early seral stages by a single event such as a fire;

(4) Spatial autoecological requirements – the prioritization of areas should also satisfy species' spatial autoecological requirements. Ideally, a conservation area network should represent at least the minimum population size required for the persistence of each species, but methods for calculating this value have not yet been extended to large numbers of species (Beissinger 2002) – see Section 6.2 below. Many species will have particular requirements for the configuration of conservation areas. These include altitudinal migrants (Powell & Bjork 1995) and species requiring particular combinations of habitat in single conservation areas (Kerley *et al.* 2003);

(5) Source–sink population structures – when species have a source–sink population structure in which a small percentage of suitable habitat provides most of the recruits for other areas (Pulliam 1988), the source habitats should be assigned high priority for conservation. Areas containing sink habitats (for instance, anthropogenically transformed areas for many species) should be excluded from consideration when assessing the representation of such species in a conservation area network. It is possible to drive area selection algorithms to select source areas if this is seen to be desirable and if suitable data are available (Nicholls 1998);

(6) Effects of habitat modification – conservation areas in fragmented landscapes require special management to safeguard surrogate persistence, such as habitat restoration and the addition of new habitat along the perimeter of the conservation area network (Saunders *et al.* 1991; White *et al.* 2004; Lee & Thompson 2005; Tabarelli & Gascon 2005);

(7) Species as evolutionary units – prioritization should give preferences to areas with physical properties believed to encourage speciation (such as the interface between soil types) or areas containing taxonomically distinct species or species with radiating phylogenies (Fjeldsa 1994; Pressey *et al.* 2003; Rouget *et al.* 2003a).

These principles all underscore that the mere representation of biodiversity surrogates does not suffice for conservation planning. Planning for persistence requires, at the very least, incorporation of rules of spatial configuration that take these seven sets of ideas into account. These rules can be incorporated during the selection of areas, either using

trade-offs or using configuration rules to break ties, for instance, by giving a preference to adjacent areas, as discussed in Chapter 5 (Section 5.5).
Alternatively, they can be implemented using multi-criteria analysis as
will be discussed in Chapter 7.

6.2 Viability analysis

Methods for assessing vulnerability include a suite of techniques that have
come to be known as "viability analysis." In the United States, the 1976
National Forest Management Act (NFMA) required the United States
Forest Service to "provide for diversity of plant and animal communities based on the suitability and capability of the specific land area"
(Sarkar 2004). In 1979 the planning regulations developed to implement
this provision required the Forest Service to "maintain viable populations of existing native and desired non-native vertebrate species in the
planning area" (Sarkar 2004). This provision was interpreted as requiring the analysis of viability of populations and led to what is called
"viability analysis" in North America (and is known by a variety of names,
including "vulnerability analysis," elsewhere). The early attempts to use
island biogeography theory to determine the prognosis for biodiversity
at a place from the extent of its area (see Section 6.1) constituted a rudi-
mentary viability analysis based on community ecology. As noted earlier
(Section 6.1) these attempts were abandoned on grounds of implausibil-
ity. Population viability analysis (PVA; see Section 6.2.1) attempts to do
better by drawing, instead, on population ecology.

6.2.1 Population viability analysis

In population viability analysis (PVA), the ideas and techniques of population ecology are brought to bear on the problem of assessing the
likelihood of the persistence of populations (Boyce 1992). PVA can be
applied to both small and large populations. When the focus of interest is on threatened and endangered species with small populations, PVA
requires the analysis of stochastic models because small populations have a
high probability of becoming extinct due to chance fluctuations. Shaffer
(1978) provided a systematic framework for such an analysis and introduced the once influential idea of a minimum viable population (MVP),
the definition of which involved choices about the probability of persistence that is sufficient as a conservation goal, and the time up to which
this probability has to be maintained. One common choice was to define

an MVP as a population that has a 95 % probability of surviving for the next 100 years (Shaffer 1978; Hanski 1999). Neither of these numbers has a firm biological basis, although for the concept of the MVP to be operationalized some such precise numbers were necessary.

By the late 1980s, it became clear that the concept of an MVP was of limited use. Even for a single species, populations in slightly different habitat patches may show highly variable demographic trends, especially in the presence of stochasticity, resulting in high variability of estimated MVPs depending critically on local context (Margules 1989). Since then PVA has focused on estimating more robust parameters such as the expected time to extinction. There have been many models of stochastic PVA incorporating demographic and environmental stochasticity (Lande *et al.* 2003).

Caughley (1994) argued for an emphasis on declining populations and especially on diagnosing the cause, reversing the decline and establishing theoretical generalities about the processes of extinction. Generally, the analysis of small populations has contributed little to the conservation of species in the wild. Beyond the relatively trivial insight that small populations are subject to dangerous chance fluctuations, stochastic models do not provide insight into why species are at risk. Conservation does have a better chance of success with large populations because it is usually possible to design field experiments with appropriate controls to determine the ecological mechanisms of decline. (For a response to this argument, see Hedrick *et al.* [1996].) However, both small and large populations can decline and the while detection of a decline before a population becomes small is preferred, smallness itself is not necessarily the problem. It is the failure to understand the cause of the decline and the processes involved that is often the real problem. Boyce (1992) has similarly emphasized the importance of modeling the interactions responsible for population declines. Chapter 4 of Caughley and Gunn (1996) details 17 case histories of species with declining populations, stressing that an understanding of the causes of decline was necessary if recovery was to occur.

Only the simplest models for population viability can be analyzed (for examples, see Lande *et al.* 2003). However, plausible models for application in the field require so much life history and other ecological detail that the models can only be explored by simulation. This limitation applies to most models for PVA that incorporate any of the seven sets of ideas discussed in Section 6.1. Partly as a consequence of this, no generally accepted unified framework for PVA has yet emerged

(Beissinger 2002). Several software packages are available for PVA including ALEX (Possingham and Davies 1995), GAPPS (Harris *et al.* 1986), MARK (White and Burnham 1999), PATCH (Schumaker 1998), RAMAS (Schultz *et al.* 1999) and VORTEX (Lacy 1993). Reviews of these software packages include Lindenmayer *et al.* (1990) and Brook *et al.* (2000).

Example 6.1   Population viability analysis of Leadbeater's possum in Victoria, Australia

An endangered species, Leadbeater's possum, *Gymnobelideus leadbeateri*, is a small nocturnal marsupial with a limited distribution restricted to stands of mountain ash (Eucalyptus regnans) forest within a 4800 km^2 region in the Central Highlands of Victoria (south-eastern Australia). The main threat to the survival of the species comes from the disappearance of nest sites in trees that are several hundred years old. A major factor in decreasing available habitats were wildfires in 1939 which burned 70 % of the mountain ash forests of the region. In the late 1980s and early 1990s, potential nest sites were decreasing by about 4 % per year (Lindenmayer *et al.* 1990). Lindenmayer and Possingham (1996) used ALEX to evaluate the effects of different forest management strategies on the survival of *G. leadbeateri*.

The study evaluated the risk of metapopulation extinction as a function of: (1) the number and spatial configuration of forest patches that are protected from logging; and (2) the impacts of post-fire salvage logging in the protected areas. Nineteen life history attributes gleaned from previous experimental studies were used in the simulations (see Table 6.1). The simulations assumed a metapopulation structure; only the females were modeled (because they are the limiting sex in the demography of such sexually reproducing species). Three age classes (newborn, <1 year; subadults, between 1 and 2 years; and adults, >2 years) were modeled. It was assumed that individuals did not breed successfully on the average in one of six years. Environmental variation and demographic stochasticity were included in the models. Figure 6.1 shows the stages of the simulation.

Models of complex patch structures within two blocks (Murrindindi and Steavenson) included three types of forest: (1) stands of old growth; (2) forests within 20 m of a stream and stands on steep and rocky areas (>30°); and (3) the remaining part, which was designated for timber

Table 6.1 *Life history attributes used in a PVA of* G. leadbeateri[a]

Attribute	Value
Minimum home range of females in highest quality habitat (which sets the limit for carrying capacity)	1.0 ha
Minimum home range size of breeding females in old growth (calculated by ALEX)	3.3 ha
Maximum population density (females/ha)	2
Reproduction	
Annual probability of 0 female young per female	0.45
Annual probability of 1 female young per female	0.30
Annual probability of 2 female young per female	0.18
Annual probability of 3 female young per female	0.06
Annual probability of 4 female young per female	0.01
Age at sexual maturity (years)	2
Mortality (annual probability of death)	
Newborn	0.0
Juvenile	0.3
Adult	0.3
Population Growth	
Population growth rate under ideal conditions (calculated by ALEX)	1.21
Population threshold for quasi-extinction	2
Movement	
Mean migration distance of juveniles	2 km
Population density before migration (percentage of maximum)	20%
Migration probability of subadults	0.7
Population density before diffusion (percentage of maximum)	10%
Diffusion probability of subadults	0.2

[a] From Lindenmayer & Possingham (1996). Reproduced with permission from Blackwell Publishing.

harvesting. Data on the distribution of these types were extracted from a GIS model developed independently by a governmental natural resource agency. Pairs of patches less than 200 m apart were assumed to be connected by a "corridor" permitting migration. Each 1 ha patch was assigned a habitat suitability index between zero and one based on ecological studies; this value was used to determine the maximum number of breeding adult females per ha. Submodels were used to track the dynamics of habitat suitability over time. ALEX includes migration and diffusion submodels to simulate the spatial movement of species.

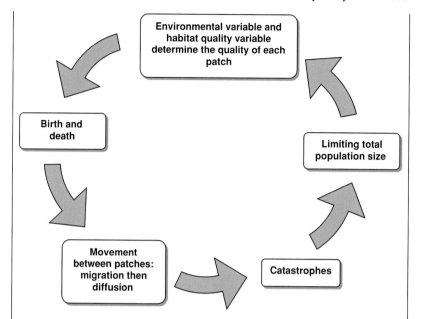

Figure 6.1. Stages of the simulation of population change in Leadbeater's possum. (From Possingham & Davies (1995).)

Both processes were included in the analysis. The simulations also included wildfires and various management options.

Five scenarios were examined in detail: one was a baseline scenario in which current logging practices were followed and the annual probability of wildfire was set at 0.01 with either 50% or 75% of patches being burnt during each wildfire event; two scenarios, one for each block (Murrindindi and Steavenson) in which 300 ha of forest were set aside for conservation in different ways (from a single 300 ha conservation area to 12 25 ha areas); and two similar scenarios in which 20 areas, each of 50 ha, were set aside for conservation.

For the first scenario, in which current logging practices were continued, the models predicted a high probability of extinction. The median time to extinction was ten years. The second and third scenarios showed that exclusion of salvage logging operations from burned old growth forests significantly improved the species' probability of survival both in the short and long term. Similarly, exclusion of logging from the identified forest patches also increased these probabilities. However, there would be a delay of at least 150 years until areas protected now made a significant contribution to the probability of

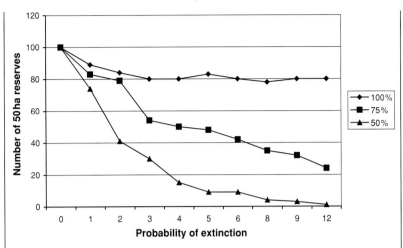

Figure 6.2. PVA results for Leadbeater's possum showing the predicted probabilities of extinction as a function of the number of 50 ha conservation areas set aside. Three different fire regimes are shown with the annual probability of fire set at 0.01. (From Lindenmayer & Possingham (1996).)

persistence of metapopulations because this is the time required for existing stands of regrowth to develop the old gowth characteristics, including hollows, that provide essential habitat to *G. leadbeateri*. Populations in smaller conservation areas were vulnerable to extinction due to demographic and environmental stochasticity. A small number of large conservation areas led to a high vulnerability to extinction due to a single catastrophic wildfire. The best design was one with a dispersed network of conservation areas (see Figure 6.2, which is based on the fourth and fifth scenarios). Sensitivity analyses with varying frequency and extent of wildfires showed that the best management strategy was to set aside several 50–100 ha conservation areas in each forest block and to exclude post-fire salvage logging operations in the event of a wildfire.

PVAs such as that of Example 6.1 can only be performed for single species or a very few species at a time. Moreover, the level of detailed knowledge of parameters required (see Table 6.1) restricts such analyses to the few charismatic well-studied species for which such data is available (for instance, *G. leadbeateri* and the northern Spotted Owl in the United States, *Strix occidentalis caurina* (Forsman *et al.* 1996)). When the prognosis for hundreds of species must be simultaneously determined, and details of life history parameters are typically unknown, PVA is effectively useless. This is the typical situation in systematic conservation planning. However, this criticism should not be misinterpreted to suggest that PVA has no role to play in conservation planning. When conservation plans have to be devised for individual species of obvious importance – for instance, phylogenetically uncommon or endangered species – adequate PVAs may well be critical for planning purposes.

However, even in such cases, technical limitations of PVA pose problems. Models for PVA have proven to be notoriously susceptible to two problems: (1) *partial observability* – even simple parameters such as the intrinsic growth rate of a population or the carrying capacity of a habitat are often intractably difficult to estimate with requisite precision from field data (Shrader-Frechette & McCoy 1993) and (2) *structural uncertainty* – slight changes in the values of parameters and apparently slightly different models of the same biological situation make radically different predictions (Burgman *et al.* 1993). This phenomenon has been observed for almost all simulation and analytic models used for PVA.

One solution has been to suggest that PVA should not be used to make absolute predictions about parameters such as the expected time to extinction. Rather, it should be used to rank different management plans for the same population, but only when all models produce the same ranking (Boyce 1992; Possingham *et al.* 1993; McCarthy *et al.* 2003). Alternatively, it can similarly be used to rank different populations under the same habitat management plan, once again when all models produce the same ranking (Sarkar 2004). From this point of view, PVA should be used neither to estimate parameters such as the expected time to extinction or to quantify viability. Rather, they should be restricted to the comparative assessment of management options. (For reviews of issues connected to PVA, and a large number of examples illustrative of the variety of techniques used, see Brook *et al.* (2000), Beissinger and McCullough (2002), Morris & Doak (2002), and Reed *et al.* (2002).)

6.2.2 Habitat-based viability analysis

As noted earlier, PVA of the type discussed in Section 6.2.1 typically cannot be carried out for more than a few species at a time because of the difficulty of obtaining sufficient data to characterize population demographics, structure, etc., with sufficient precision to make estimates of parameters such as expected times to extinction sufficiently reliable for use in conservation planning. An alternative promising method is to attempt to assess viability using the ecological requirements of species (Boyce *et al.* 1994). This method, which is similar in spirit to the modeling techniques discussed in Chapter 4, has been used with some success for large animal species and involves the inference of habitat requirements through studies of use versus availability of resources. All such models are constrained by both statistical and ecological limitations (Garshelis 2000).

Techniques for habitat-based viability analyses include the use of both systematic simulation (Akçakaya & Atwood 1997) and statistical models (Manly *et al.* 2001). The latter are useful when the problem is that of predicting the persistence of many species simultaneously as is typically encountered in conservation planning. In particular, resource selection functions (RSF) have emerged as a promising statistical modeling technique (Alldredge & Ratti 1992). This method relies on the measurement of the use of resources by animals relative to resource availability. A positive relation suggests selection, a negative one avoidance. Animal relocation data can be used to estimate the strength of resource selection. This leads to the formulation of predictive equations that allow extrapolation of these relationships across landscapes. Resource selection by animals occurs at different spatial scales, with different resources being selected at each scale. The predictive equations are hierarchical, incorporating the interactions at each scale. Thus resource selection models are similar to the niche models discussed in Chapter 4, insofar as their direct purpose is to predict the distribution of species. However, when these models are used in conjunction with PVA (as in Example 6.2), we have a promising way to integrate spatial knowledge of distributions with viability analysis.

Example 6.2 Habitat-based population viability analysis of wildlife of the Canadian central Arctic (Plate 19)

Johnson and Boyce (2005) carried out a habitat-based PVA for wildlife of the Canadian central Arctic. Recent discoveries of mineral deposits across the region had led to unprecedented levels of mineral exploration and development. The cumulative effects of these activities, as well as those of recreation and other mineral and road development, had led to concern among resource managers and others (including First Nation representatives) about the future of the region's biota. Johnson and Boyce focused on the statistical relationship of four animal species (barren-ground caribou, *Rangifer tarandus groenlandicus*; wolves, *Canis lupus*; grizzly bears, *Ursus arctos*; and wolverines, *Gulo gulo*) to vegetation, interspecific interactions and human disturbance over a large study area. The caribou is a major prey of the three carnivore species.

Figure 6.3 shows the protocol used. The analysis consisted of three interrelated steps and integrated RSF analysis with a spatially explicit PVA for two of the species (see below).

Stage 1
Scoping began by identifying the geographical area (191 000 km^2; see Figure 6.4), which was centered at Contwoyto Lake (65° 30′ N, 110° 30′ W) about 400 km north-east of Yellowknife, Northwest Territories (Canada), and consisting of the Taiga Shield and Southern Arctic ecozone. The four species mentioned above were chosen as focal species and the ecological relationships between these species, resources and disturbance effects were identified. Anthropogenic factors operating in the region include subsistence use and occupancy by First Nation communities, industrial mining operations dating back to the 1930s, and associated infrastructures. The analysis was based on information on animal captures and relocations from past work which used satellite radio-collars to track movements. Table 6.2 shows the resource and human disturbance variables used to model resource selection. Vegetation habitat variables were derived from a supervised classification of (LANDSAT) remote-sensed satellite data. The disturbance factors that were identified were buffered to include larger areas as a precautionary measure to include the full effects of disturbances.

Stage 2
The RSF analysis identified the habitat requirements of the selected species. RSFs quantify the relationship between the observed distribution of species and covariates representative of habitats and human disturbance. Typically, an RSF consists of a number of coefficients that quantify selection for or avoidance of environmental features. The analysis used conditional fixed-effects logistic regression to estimate coefficients and an information-theoretic approach, based on the Akaike information criterion (AIC) (Burnham & Anderson 1998) to guide model development and selection. This allowed the choice of the model with the greatest explanatory power while minimizing bias and maximizing precision of the model parameters. The predicted

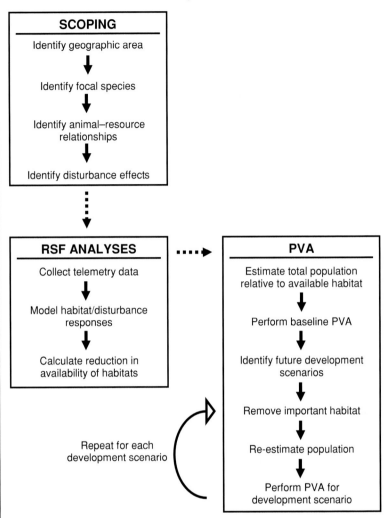

Figure 6.3. The analytic framework used for a habitat-based PVA in the Canadian Arctic. The approach began by "scoping" the study bounds and identifying ecological relationships of interest, generating resource selection functions for selected species, and using population viability analyses to assess demographic effects of developments. (From Johnson & Boyce (2005).)

spatial distributions of habitats for each species were then calculated for each season for three scenarios, each representing a management option: (1) a "present" or "modeled disturbance" scenario based on current maps; (2) a "no disturbance" scenario; and (3) an "assumed disturbance" scenario projecting future effects on the habitat because

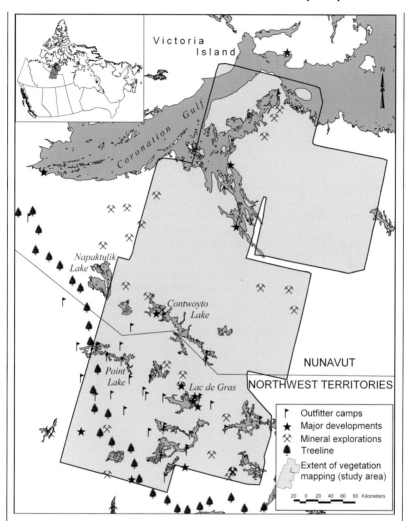

Figure 6.4. Habitat-based viability analysis in the Canadian Arctic. The study area across the Canadian central Arctic with the location of disturbance features. The extent of the study area was determined by the availability of vegetation mapping. The treeline represents the northern limit of coniferous forest. (From Johnson & Boyce (2005).)

Table 6.2 *Independent variables used to model resource selection by caribou, wolves, grizzly bears and wolverines across the Canadian central Arctic[a]*

Variable	Description
Resource Variables	
Esker density/patch[b]	Sparsely vegetated sand and gravel esker complexes
Forest density/patch	Continuous or discontinuous forested areas of dwarf white spruce, black spruce and tamarack
Heath-rock density/patch	Open mat heath tundra interspersed with bedrock and boulders
Heath-tundra density/patch	Closed mat of heath found on moderate-to well-drained soils on upland areas
Lichen-veneer density/patch	Windswept, dry, flat topography covered with a continuous mat of lichen
Low-shrub density/patch	Extensive areas of low birch and willow found on moist well-drained soils
Peat-bog density/patch	Mosaic of uplands and lowlands with fens, bogs, mixed-wood forest and peatlands
Riparian-shrub density/patch	Active stream courses or areas of water seepage with a shrub layer of birch, willow and alder
Rock-association density/patch	Large areas of windswept bedrock or boulders with little vegetative cover
Sedge-association density/patch	Wetland complexes of wet sedge meadow and drier hummock sites
Occurrence of caribou	Predicted likelihood of encountering caribou
Occurrence of grizzly bear	Predicted likelihood of encountering a grizzly bear
Occurrence of wolf	Predicted likelihood of encountering wolves
Occurrence of wolverines	Predicted likelihood of encountering wolverines
Disturbance Factors	
Major developments[c]	Operating mines, communities, winter road camps
Mineral explorations	Areas of mineral exploration activities
Outfitter camps	Seasonal guide-outfitter camps

[a] Modified from Johnson & Boyce (2005). Reproduced with permission from Blackwell Publishing.
[b] Vegetation was modeled at two scales: the % area per type and the mean density representative of the regional distribution of each vegetation type.
[c] Disturbance was quantified as the distance of animal and random locations from the nearest facility operating during that time; non-linear effects were represented as a two-term Gaussian function.

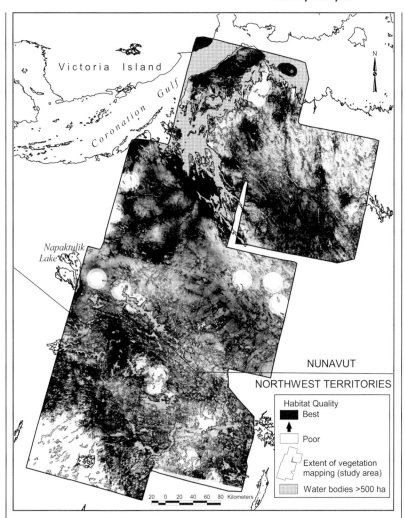

Figure 6.5. RSF-modeled grizzly bear habitat in the Canadian central Arctic. Predicted habitats for grizzly bears averaged across seasonal RSF models representing the scenario of present disturbance. Percentiles were used to rank habitats within ten classes. (From Johnson & Boyce (2005).)

of disturbance. Figure 6.5 shows the map for grizzly bears averaged over the seasons under the "present" disturbance scenario.

Stage 3
This stage consisted of the use of a spatially explicit PVA to evaluate different future scenarios corresponding to available management

options. RSF analyses are static, only indicating habitat suitability for species but not their dynamic response. To assess the prognosis for the species Johnson and Boyce (2005) supplemented the RSF analysis with a spatially explicit PVA using the PATCH software package (Schumaker 1998). This analysis was restricted to grizzly bears and wolverines, the only two species for which sufficient demographic data were available.

The purpose of the PVA, as in Example 6.1, was to assess management options. PATCH simulates the dynamics of populations of females using their survival and fecundity within specific habitat types. The input consists of a number of parameters, which PATCH uses within its simulation routines to track the births, deaths and movements of individual animals through time and between territories, growing, extirpating or holding constant an initial seed population. Variation in demographic performance is a function of age-specific survival and fecundity summarized in a population projection matrix (the Leslie matrix) and scaled across habitat classes that vary in importance to the study species. Demographic simulations were conducted for a 250-year period: populations were subjected to past disturbance (1890–1990) for the first 100 years, present levels of disturbance for the next 50 years (1991–2041), and future levels of disturbance (2042–2141) for the last 100 years. The 250-year projection was run four times, first incorporating disturbance and habitat loss, but no environmental variance, and then using vital rates sampled from one of three distributions defined by progressively larger variances (SD = 0.025, 0.075, 0.125). Five hundred replications were used. The variance measured the level of environmental stochasticity.

Results illustrated the mean (± 95 % confidence intervals) population trajectories across time for each level of environmental variance, the percentage of simulated populations that reached an arbitrary quasi-extinction threshold of a 90 % reduction in adult females, and the spatial distribution of source–sink territories.

Results and implications
The results of the RSF analysis showed that, across all the models, mines and other major developments had the largest negative effect on species' distributions, followed by exploration activities and outfitter camps. Wolves and grizzly bears were most impacted by human disturbance throughout the year. Caribou showed the strongest seasonal response; there was a 37 % decrease in high-quality post-calving

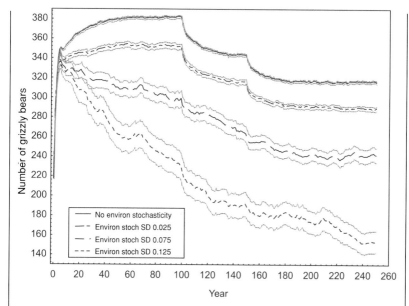

Figure 6.6. PVA results for grizzly bears in the Canadian central Arctic. Mean (±95 % confidence intervals) predicted number of female grizzly bears for simulated populations of the Canadian central Arctic. Demographic models were developed for four levels of environmental stochasticity; disturbance-related habitat reductions occurred at year 100 and 150. (From Johnson & Boyce (2005).)

habitats. The analysis did not support a strong avoidance response to all sources of disturbance for any of the modeled species.

Confirming what we said in Section 6.2.1 about the paucity of adequate data, Johnson and Boyce had to restrict the PVA to only two of the four species studied. Figure 6.6 shows typical results of the PVA in the case of grizzly bears. Grizzly bears showed a decline due to the effects of present and future disturbances; wolverines showed little sensitivity to these factors. Only the former can plausibly be regarded as vulnerable.

It is important to emphasize that RSF-based methods are relatively new and must be used with caution. Lawler and Schumaker (2004) report negative results in an attempt to relate simple habitat preference measures to the distribution of red-shouldered hawks (*Buteo lineatus*) and northern goshawks (*Accipiter gentilis*) in the mid-Atlantic region of the United States. In general, habitat-based viability analyses are likely to make optimistic predictions of suitability of habitats and occupancy because they typically (though not always, see Akçakaya [2000]) ignore the spatial metapopulation structure of many species. Consequently they tend to identify larger areas as species' habitats than those that are actually occupied by the species.

6.2.3 Decision analysis

Given the difficulties faced in carrying out formal viability analyses as described above, less detailed, but nevertheless indispensable risk assessment and decision analysis methodologies have begun to be used to assess the prognosis for the full complement of biodiversity at a place. These methods have been classified into three categories (Durant & Mace 1994; Gaston *et al.* 2002):

(1) *Subjective assessments* – these comprise the majority of decisions including many listings of taxa believed to face a high risk of extinction (for instance, in the Red data books and associated publications (see Burton [1984] and Munton [1987] for examples). They represent the subjective judgements of individual researchers or groups of researchers reflecting their experience of the taxa or habitats. They are often valuable in providing broad guidelines about spatial and temporal factors that should be taken into consideration for adequate conservation planning;

(2) *Heuristic rules* – these are supposed to be an advance over purely subjective assessments. These rules include the use of a small set of life history or population parameters to distinguish taxa at different risk levels (e.g., East 1988, 1989; Millsap 1990; Molloy & Davis 1992; Reed 1992; Dickman *et al.* 1993; Given & Norton 1993). Most such methods assume that taxa with small populations or small ranges have an increased risk of extinction without paying much explicit attention to temporal demographic trends or distribution patterns. Typically, they ignore genetic factors as well as the potential for extreme environmental stochasticity (the occurrence of catastrophes). Consequently, it is possible that they do not identify taxa that may be at risk because of low density in spite of large geographical range (McIntyre 1992; Gaston 1994) or populations that are at risk because a large fraction of individuals occur in sink rather than source populations (Pulliam 1988; Dias 1996). Heuristic rules used in the past include the (in)famous 50/500 rule for the size of an MVP, with 50 supposed

to be the MVP size in the face of demographic stochasticity and 500 the MVP size in the face of environmental stochasticity (see Simberloff [1988]; Boyce [1992]; and Caughley [1994] for further discussion). The most that can be hoped for from such heuristic rules is the identification of those taxa that should be of highest concern during conservation planning;

(3) *Quantitative decision analysis*, which includes traditional and habitat-based viability analysis as a partial case of decision analysis in which risk is quantitatively assessed but the cost or utility of a decision, one management option versus another, is not. However, most decision analysis protocols do not require a full PVA for risk assessment. We will discuss quantitative decision analysis in some detail in Section 6.4 below because we anticipate its increasing use in systematic conservation planning.

All three of these methodologies can be applied to individual taxa (obviously including species), sets of taxa and, most importantly, to entire habitats.

6.3   Targets for representation

We mentioned in Chapter 5 (Section 5.2) that explicit targets for representation must be introduced for a quantitative solution to the problem of economical representation of biodiversity surrogates in a nominal conservation area network. These targets must be explicit and encompass both patterns (or distributions of typical biodiversity surrogates) and processes. Various targets have been suggested in the literature including 10%, 12%, 15% and even higher values. However, none of these numbers is based on solid biological grounds. Moreover, such uniform targets over all biodiversity features are unlikely to capture what is individually required by each surrogate for its continued persistence.

Recall that representativeness was a central goal of conservation planning. The importance of setting explicit targets is that representativeness can only be non-arbitrarily assessed provided that an explicit target of representation is set for each surrogate. For patterns such as the distribution of biodiversity surrogates, these explicit targets should also be quantitative (either a certain number of times or a certain fraction of its occurrence in the data set). Representativeness can then be viewed as measured by the fraction of attributes that meet their set targets. Moreover, having explicit targets also permits an assessment of the degree to which a second goal of conservation planning, economy, is being achieved in a nominal conservation-area network. Recall Soulé and Sanjayan's (1998) argument that targets such as 10% and 12% are far too small for the persistence of many species and have only been adopted for political expediency. The only adequate response to this argument is to show how targets can be chosen to meet the relevant needs of the biological entity of interest. Thus targets must be designed to ensure the third goal of conservation planning: the persistence of species and other biodiversity surrogates.

There is no accepted framework for doing this in all cases, but setting targets with persistence in mind is not entirely arbitrary. We will illustrate how biological principles can be used to set targets using two examples. The first of these involves the use of heuristic rules based on planners' experience of the study region. The second uses generic reasoning about population viability with sparse data to set targets of representation for plant species. Similarly, RSF analysis, as described earlier (Section 6.2.2), can potentially be used to set targets for animal species. An important point that deserves emphasis is that, if persistence is the goal, targets should not be set using considerations of what is feasible because of socio-political and other similar constraints (Pressey *et al.* 2003). Though these other criteria are undeniably critical to conservation planning (see Chapter 7), their presence may require us to accept that certain biodiversity targets cannot be satisfied. This may, in some circumstances, enable the revision of these other socio-economic goals. What would be undesirable is an unwarranted sense of confidence in the persistence of biodiversity when non-biological factors have been used to set conservation targets, the very thing that concerned Soulé and Sanjayan (1998).

Example 6.3   Formulating targets for the Cape Floristic Region of South Africa

In Chapter 8 (Section 8.4), we will describe a conservation plan produced for the Cape Floristic Region of South Africa (Plates 17 and 18). Here we discuss how targets were set for both biodiversity patterns and processes (Pressey *et al.* 2003). Table 6.3 lists the features for which targets were set during this exercise.

Broad habitat units
Cowling and Heijnis (2001) mapped broad habitat units (BHUs) for the region using climatic zones and vegetation types resulting in

Table 6.3 *Conservation features used to set targets for the Cape Floristic Region. See text for detailed discussion*[a]

Features	Number of features	Target units
Land types		
Broad habitat units	102	Ha
Species		
Proteaceae species	364	No. of records
Vertebrate species	365	No. of records
Large and medium-sized mammals	41	No. of individuals
Processes		
Edaphic interfaces	8	km of interface
Upland-lowland interfaces	146	km of interface
Sand movement corridors	6	Whole corridors
Inter-basin riverine corridors	6	Whole corridors
Upland–lowland gradients	85	Whole gradients
Macroclimatic gradients	3	Whole gradients

[a] From Pressey *et al.* (2003). Reproduced with permission from Elsevier.

102 types for the region. Targets were set for the representation of these based on two premises: (1) the required percentage target increased with physical or biological heterogeneity, rarity and vulnerability; and (2) the percentage target was based on the original extent, preceding intensive land use, rather than current extent, resulting in larger targets and decoupling these from ongoing habitat transformation.

First, a baseline target was set based on heterogeneity which was measured by inferred relative magnitudes of species turnover. The following generalizations were used for this inference: areas in the western mountain subregion have more than double the species richness compared to the eastern subregion, western lowland areas have 1.5 times the species richness compared to the eastern subregion, and patterns of richness are driven by turnover of rare plant species. These ideas were incorporated into the formula for the baseline target, B:

$$B = b \times \frac{A}{100} \text{ ha} \tag{6.1}$$

where A was the total area of the BHU, b was ten for lowland and mountain BHUs in the eastern subregion, 15 for lowland BHUs in the western subregion and 25 for montane BHUs in the western subregion. B varied from 180 ha to 93 000 ha. If the targets were larger than the current extent of a type, restoration was envisioned to achieve the required level of representation.

Second, retention targets were used to address threat to the extant vegetation. BHUs were classified into three threat categories (high, H; moderate, M and low, L) and three threatening factors (agriculture, alien plants and urbanization). Explicit rules were formulated to assign each BHU to one of the three threat categories. For instance, for alien plants, if $H + M > 15\%$ of the remaining area of the BHU, then threat was assigned as H; if $1\%\ H + M \leq 15\%$, threat was assigned as M; otherwise it was L. The overall threat category of a BHU was its highest one according to any of the three factors. Retention targets, R, were then allocated according to the formula:

$$R = t \times \frac{A}{100}\ \text{ha} \qquad (6.2)$$

where A was the total area of the BHU, t was 30, 15 or 0 for threats H, M or L, respectively. Thus retention targets increased in size with the level of threat though the numbers used (30, 15 and 0) remain arbitrary. R varied from 0 ha to 129 000 ha.

Third, an adjusted target, Ta, was calculated on the assumption that the targets were not achieved mainly by including areas in edaphic and upland–lowland interfaces. This is intended to ensure that both marginal and core areas of land types are included in a conservation-area network. Ta was calculated by the formula:

$$Ta = Tp \times \frac{I + 50}{100}\ \text{ha} \qquad (6.3)$$

where the provisional target, Tp, is equal to $B + R$, I is the percentage of Tp within interfaces. This formula ensures that at least 50% (an arbitrary choice) of the areas selected would be outside the interfaces.

The final targets for BHUs were the sum of these three targets.

Species

Proteaceae species   Once again baseline and retention targets were set. The baseline target was ten occurrences for non-sprouting species with less than 50 records in the region and five for all other species. The rationale for a higher target for non-sprouting species was that these are subject to fire-induced local extinction; however, the precise numbers remain arbitrary. Retention targets were set by a method similar to the one described above. First, every record was placed in a threat category, *H*, *M* or *L*, on the basis of each of the three factors mentioned earlier, with the final assigned threat category being the maximum. This allowed the creation of a threat profile for each species with its records being vulnerable (*H* or *M*) or non-vulnerable (*L*). The retention targets were then set to ensure, where possible, that at least five records, in addition to baseline targets, were non-vulnerable. The final targets (baseline + retention) were truncated to the total number of records when necessary.

Vertebrate species   A baseline target of one and a retention target of zero or one based on vulnerability was used.

Large and medium-sized mammals   Targets were set for these 41 species (with average body size higher than 2 kg) using explicit rules. First, records of species in vegetation fragments that were presumed: (1) not to be able to hold populations that were of functional size or (2) not capable of being maintained by conservation agencies were removed from consideration. Then, the following rules were used:

(1) Species with only marginal historical occurrences in the Cape Floristic Region and not listed as threatened were set targets of zero (three species). The assumption was that the onus for the protection of these species falls on other regions;
(2) Species that are endemic or near-endemic to the region had targets of 2000 individuals assigned to them (three species). This target is supposed to represent the minimum size required to maintain evolutionary processes such as mutation (Lacy 1997). The assumption was that the persistence of these species is largely dependent on processes within the region;
(3) For species of large carnivore with substantial populations in the region but also distributed elsewhere (three species: brown hyena, *Hyaena brunnea*; spotted hyena, *Crocuta crocuta*; lion, *Panthera leo*) targets were set at 50, using the 50/500 rule for persistence in the face of demographic stochasticity (see Section 6.2.3);
(4) For one species of carnivore, the African wild dog, believed to occur in a large metapopulation of tiny populations, the target was set at ten following the South African national conservation strategy (Woodroffe & Ginsberg 1997);

(5) All other species had targets set to 200 individuals, which was supposed to be the minimum size necessary to avoid inbreeding depression (Caughley 1994).

Processes   First, the spatial scale of various processes was categorized into five classes. Targets were then set for the following six spatial surrogates for these processes.

Edaphic interfaces   These are specific juxtapositions of soil types supposed to encourage evolutionary diversification on the two sides. For the eight such interfaces occurring in the region, the entire untransformed interface with a 500 m buffer on either side was targeted for conservation; transformed sections were candidates for restoration.

Upland–lowland interfaces   These differentiate elevational and associated climatic parameters and were supposed to aid ecological diversification of plant lineages and facilitate seasonal faunal movements. Once again using a 500 m buffer on either side, the entire untransformed sections of 146 types of interface were targeted for conservation; transformed sections were candidates for restoration.

Sand movement corridors   Seven major sand movement corridors were identified in three coastal BHUs. These historically provided gradients of soil development important for soil-specific plant assemblages and diversification in some plant lineages (McLachlan & Burns 1992). Such a corridor was deemed functional if less than 50 % of each section (source, core, downwind) had been transformed. Six targeted corridors were selected comprising about 19 500 ha of which 80 % is untransformed and another 15 % likely to be restorable.

Interbasin riverine corridors   Interbasin rivers were supposed to facilitate dispersal of biota between different environments. Six of these were identified; areas selected for conservation included all second-order and the longest first-order tributaries with all streams buffered by 250 m on each side. The total riverine corridor length was 6700 km.

Upland–lowland gradients   These were identified as 1 km wide pathways that extended across adjacent lowland and upland areas of the BHUs. These were supposed to be associated with diversification of animal lineage differentiation. There were 85 upland–lowland gradients of three types. For each of them two sets of paths of 1 km width in each direction were selected. These paths were supposed to enhance the connectivity of the network.

Macroclimatic gradients   Protecting these was supposed to aid adjustment of species' ranges to climate change and also facilitate dispersal of biota between conservation areas. Three such gradients connecting upland biogeographic zones were selected. As in the case of upland–lowland gradients, three 1 km wide paths were selected. Gradients in the lowlands were not similarly targeted because there was less evidence for their diversification.

Example 6.4   Setting targets for the size of plant conservation areas

Burgman *et al.* (2001) developed a target-setting method that can be rapidly used to determine the size of conservation areas for vascular plants. The method is based on the assumption that the risk of decline or extinction that all populations face because of accidents of natural temporal and spatial variation can be modeled by simple population models that include demographic and environmental stochasticity. We will describe the method here and illustrate it with one of their examples, *Banksia cuneata*, an endangered shrub that grows up to 5 m in size in six localized stands in the undulating sand plains of south-western Australia.

There are twelve steps:

Step 1
The method begins by estimating the population size, F, likely to persist under the effects of demographic and environmental stochasticity in the absence of anthropogenic factors. Burgman *et al.* (2001) define F as the size of a population with less than a 0.001 (0.1 %) probability of falling below 50 adults at least once in 50 years.

As noted earlier, 50 years provides a short-term scale adequate for planning purposes. F can be estimated from knowledge of the life history of a species or of similar species or using simple population models which include demographic and environmental stochasticity. Table 6.4 provides some indicative numbers.

Table 6.4 *Initial population size of* Banksia cuneata *to achieve a probability <0.01 of falling below 50 mature individuals at least once in the next 50 years. The model assumes a continuous regeneration response and average growth rate of one (no growth or decline) under deterministic conditions*[a]

| | Initial population size | | | | |
| | Coefficient of variation in vital rates | | | | |
Survivorship	0.05	0.10	0.15	0.20	0.25
0	520	1 000	7 500	23 000	60 000
0.2	480	800	2 500	17 000	50 000
0.5	390	650	1 800	12 500	44 000
0.9	280	550	1 650	9 800	40 000
0.98	180	500	1 600	6 000	38 000

[a] From Burgman *et al.* (2001). Reproduced with permission from Blackwell Publishing.

For *B. cuneata*, F was estimated as 6400 (5400, 7400) in a detailed simulation done by Burgman and Lamont (1992). The numbers in brackets represents bounds for the population.

Step 2
The second step is to identify disturbance regions, which are populations or groups of populations that experience the same type of disturbance regime as used to estimate F.

For *B. cuneata*, all potential habitat was in the same disturbance region.

Step 3
The next step is to identify and map the area of potential habitat. Such a map is necessary for each disturbance region. Expert knowledge can be used to generate such a map, as can the techniques described in Chapter 4.

For *B. cuneata*, there were six remaining populations of fewer than 400 mature plants in each. The total range was less than $60 \times 60\,km^2$.

Step 4
Next, the area of potential habitat, H_i, surveyed in each disturbance region, i, must be outlined. Here, the techniques described in Chapter 3 may be used.

For *B. cuneata*, the total area of its occurrence has been surveyed and is known in detail.

Step 5
For each disturbance region, the size of the adult population, N_i, and density, D, with $D = N_i / H_i$, must be estimated. A survey-based estimation may be difficult and it may be reasonable to use models to obtain a value for D.

Stands of *B. cuneata* are dense but occur only sporadically. Average D is about 10 plants/ha with a range of 5–15 plants/ha. The range reflects uncertainty about what constitutes suitable habitat.

Step 6
The raw target area for conservation, A_0 , may now be estimated using the relation, $A_0 = F/D$, for each disturbance region. If this area is put under a conservation plan, the probability of the population falling below 50 individuals within the next 50 years is less than 0.001.

For *B. cuneata*, $A_0 = 6400/10 = 640$ (360, 1480) ha with the upper and lower bounds calculated using both the bounds for F and the range of D.

Step 7
The relatively small-scale disturbances affecting the species' habitat from which the species recovers over the next 50 years must next be identified. The characteristics of these disturbances must then be used to determine the proportion, S, of potential habitat available to the species at any time. In many circumstances this parameter can be estimated directly.

Alternatively, the effects of disturbances can be modeled simply if it is assumed that the disturbance events are random and independently distributed across the landscape (with respect to the species' occurrences). Let n_d be the habitat recovery time (in years), the interval between the disturbance and the appearance of reproductively mature results, and n_u the interval between the disturbance and the point at which the habitat becomes unsuitable for the species. The latter parameter is relevant for species that require disturbances for their survival, for instance those that occur in early successional stages of habitats. Let $p_x (x = 1, 2, \ldots, z)$ be the proportion of the landscape disturbed by process x per year. Then the proportion of the landscape, q_u, that is undisturbed every year is given by:

$$q_u = (1 - p_1)(1 - p_2) \cdots (1 - p_z) \qquad (6.4)$$

It is now straightforward to calculate S:

$$S = q_u^{n_d} - q_u^{n_u} \qquad (6.5)$$

Finally, the target area for the species now gets modified to: $A_1 = A_0/S$. Note that this is greater than A_0 because S is less than one, and the factor by which A_1 increases ensures that the total available habitat in the presence of disturbance is that which would maintain a population of size F.

For *B. cuneata*, $n_d = 5$, and $n_u = 45$, the age at which plants lose reproductive potential. Assume that fires originating in adjacent areas are the only disturbance and that these increase the risk by 0.01. Then $S = (1 - 0.1)^5 - (1 - 0.1)^{45} = 0.58$. This gives $A_1 = 2552$ ha.

Step 8
Next the target area must be adjusted to account for deterministic trends that irreversibly affect the species' potential habitat. Let L_i ($i = 1, 2, \ldots, n$) be the rate of potential habitat loss and C_i be the proportion of habitat lost due to process i. (This requires that $\sum_{i=1}^{n} c_i = 1$.) Then, for a 50-year period, the adjusted target area must be modified to:

$$A_2 = \frac{A_1}{\sum_{i=1}^{n} (1 - L_i)^{50} + \left(1 - \sum_{i=1}^{n} c_i\right)} \qquad (6.6)$$

For *B. cuneata*, there was no known deterministic effect of this kind. Therefore, $A_2 = A_1$.

Step 9
Next the target area must be adjusted to take into account processes that permanently reduce the density of populations. Many anthropogenic changes, for instance those due to grazing of livestock, belong to this category. Let r_j ($j = 1, 2, \ldots, m$) be the proportional reduction of local density due to the j-th process. Then the targeted area must be further adjusted to:

$$A_3 = \frac{A_2}{\prod_{j=1}^{m} r_j} \qquad (6.7)$$

This is the final estimated area. For *B. cuneata*, there was no known process of this kind. Therefore, $A_3 = A_2 = A_1$. The next three steps may be used to guide the allocation of this area among different potential locations.

Step 10
This step requires the identification of potential catastrophes (events that completely eliminate a population) likely to affect the species' potential habitat, the number of populations, dispersal abilities, etc. If possible, the probabilities of each of these must be estimated. The areas selected must be sufficiently dispersed to prevent catastrophes from eliminating all populations – Burgman *et al.* (2001) do not provide a formal model for this step.

For *B. cuneata*, there was no such predicted catastrophe. Land clearing was a possibility but, because the species is legally protected, with clearing leading to substantial penalties, it was regarded as unlikely.

Step 11
Targets must next be combined across disturbance regions so that the total area targeted in the different regions together equals A_3. Suppose that there are y disturbance regions. Then the A_3^k ($k = 1, 2, \ldots, y$) may all be different because different processes impinge on different areas. Let X^k be the area selected in the k-th disturbance region. That the total area selected maintains the minimum size population can be ensured by requiring that:

$$\sum_{k=1}^{y} \frac{X^k}{A_3^k} = \frac{X^1}{A_3^1} + \frac{X^2}{A_3^2} + \frac{X^3}{A_3^3} + \cdots + \frac{X^y}{A_3^y} \geq 1 \qquad (6.8)$$

Note that this condition can be satisfied by having any X^k be bigger than A_3^k. However, what is discovered in Step 10 may dictate against such a strategy of area selection. Note that the formula given here assumes that all individuals in the different disturbance regions can be regarded as forming a single population. For *B. cuneata* there was only one disturbance region.

Step 12
Possible strategies of selecting the total area should be evaluated using whatever knowledge is available regarding the species' persistence

requirements. For instance, in most circumstances, it would not be recommended that the entire selected area consists of small patches with a concomitant high probability of transformation into unsuitable habitat. The method outlined should not be used blindly but, rather, as providing support for decisions.

The methods described in Examples 6.3 and 6.4 allow targets for representation to be set in such a way that they are: (1) explicit; (2) quantitative to the extent possible (Margules & Pressey 2000); (3) based on an explicit discussion of what they are supposed to achieve; (4) credibly and transparently based on criteria correlated with the persistence of species and other biodiversity features; and (5) not based on consideration of what is socio-politically feasible. However, note that arbitrariness in setting some of the parameter values could not be entirely avoided. The two examples do not exhaust possible ways in which targets can be set or justified, which remains a topic for further research. Such research should aim to remove arbitrariness as much as possible.

6.4   Formal decision analysis

What contributes most to the difficulty of credible assessments of vulnerability and persistence in conservation planning, especially into the mid- and long-term future, is that decisions must be made under uncertainty. (Recall the discussion of PVA in Section 6.2.1.) This situation has led to the increasing relevance of formal decision analysis in planning. We will discuss one methodology – Bayesian decision analysis – in some detail because of its generality and applicability in many practical scenarios (Ellison 1996; Wade 2000). The Bayesian approach allows the quantitative incorporation of uncertainty at every level of the planning process from parameter estimation, through the results from a model, to the context of making decisions. However, it has its critics. See Dennis (1996) for a sceptical view of the relevance of Bayesian decision analysis in ecological contexts.

There are two main methodologies within statistics: the so-called frequentist or conventional statistics and Bayesian statistics. Consider the problem of partial observability (see Section 6.2.1) which results in uncertainties in the estimation of parameters in the field. Conventional statistics is used to estimate the value of a parameter as a single number (a "point-estimate") with an associated confidence interval. It is also used to calculate a probability value, the p-value, of a precise hypothesis about the reliability of the estimate. The p-value cannot be interpreted as the probability that the parameter has the value estimated for it; rather, given that value for the parameter, the p-value is the probability of not collecting the observed data were the population to be repeatedly sampled. Similarly, a confidence interval, say at 95 %, is interpreted as saying that, were the data sets repeatedly sampled, the true value would lie within the calculated interval 95 % of the time. It does not say that there is a 95 % probability that the true value lies in the calculated interval which, in any case, is not a probability distribution.

In contrast, Bayesian estimation results in a probability distribution for the value of the parameter. Bayesian estimation of a parameter depends on a basic rule (Bayes' rule) which connects a posterior probability distribution for a parameter and a prior probability distribution for that parameter using the likelihood of the data if the prior distribution were the correct one:

$$\text{posterior distribution } (p') \propto \text{prior distribution } (p)$$
$$\times \text{likelihood } (d \,|\, \text{prior distribution})$$

where p is the parameter being estimated and d is the data that were collected. Bayes' rule can be used iteratively, with the posterior distribution from one analysis becoming the prior distribution for the next. If an interval captures α % of the probability distribution (as the area under the curve), then there is an α % probability that the value of the parameter lies within that interval. Bayesian estimation is straightforward and transparent.

The main problem with the use of Bayesian statistics is that a prior probability distribution is necessary for the use of Bayes' rule. The usual objection is that choosing a prior distribution for the value of a parameter before measurement (or, alternatively, attributing a prior probability to a hypothesis before confrontation with data) is arbitrary. While this is an important objection there are well-known strategies to mitigate its force. "Objective" Bayesians generally prefer uninformative priors: for instance, in the absence of any information whatsoever regarding the value of a parameter, a principle of indifference would suggest that all possible values be ascribed the same prior probability. "Subjective" Bayesians prefer to use whatever little knowledge may be available about the parameter to inform the selection of the prior distribution. (For more details, see Robert (2001).)

But it is possible to make a virtue out of the requirement that a prior probability distribution must be assumed in the use of Bayes' rule: it allows the incorporation of biological knowledge into the calculation of the uncertainty of a result including, especially, the results of previous experiments or sets of observations. Suppose, for instance, the size of a population has to be estimated, as is required in almost all models of population viability analysis. Suppose that several independent surveys, each with an associated uncertainty, are carried out. Conventional statistics provides no method to combine the results of these surveys to reduce the uncertainty through a composite analysis. Bayesian analysis is ideally suited to that task. Any of the surveys can be used first; that is, the order does not matter. Starting with some prior – objective Bayesians would use an uninformative prior; subjective ones would try to incorporate any hint of biological knowledge they may have – Bayes' rule is used to generate the posterior probability distribution. Then this posterior distribution can be used as the prior to analyze the result of another survey. The process can be iterated until all surveys have been used. Typically, the result will be an increasingly sharper distribution for the size of the population with a concomitant decrease in uncertainty. (For a hypothetical example, see Wade (2000).)

Perhaps the greatest utility of Bayesian methods in the context of systematic conservation planning is that they provide a rational framework for making decisions in the presence of uncertainty when there are many options to be considered (Sarkar 2005). Conventional statistics does not allow the simultaneous consideration of more than two options (Robert 2001). Bayesian decision theory requires the introduction of a loss function, which formalizes the loss connected with each decision from a set of possible alternatives. Decisions in the alternative set have probabilities associated with them and can be of various types. Generally they arise from a posterior probability distribution obtained through the use of Bayes' rule: for instance, a parameter estimate can be regarded as a decision to accept that value for the parameter; the associated probability is then obtained from the posterior probability distribution for that parameter. This probability is now multiplied by the loss function and the product, called the expected Bayesian loss, is minimized for the entire set of alternatives.

Taylor *et al.* (1996) also performed a Bayesian PVA for this population explicitly incorporating uncertainties in estimated population parameters (see Example 6.5). Goodman (2002) and Wade (2002) discuss other ways of using Bayesian methodologies in PVA.

Example 6.5   Risk classification of spectacled eiders in Alaska

Taylor *et al.* (1996) used Bayesian analysis to address the problem of the risk classification of the spectacled eider, *Somateria fischeri*. These are large-bodied sea ducks that nest along the Arctic coast of Siberia and the Chukchi and Bering Sea coasts of Alaska. In 1993 the US Fish and Wildlife Service listed the spectacled eider as threatened; what was at stake was whether its classification should be changed to "endangered." The number of pairs nesting in the Yukon–Kuskokwim delta (the southern Alaskan metapopulation) had declined from an estimated 50 000 pairs in the early 1970s to <2000 pairs in 1992. Possible threats that were identified included altered marine food supplies, toxic metal contamination (including lead poisoning), increasing predation, over-harvest and disease, though the causes of long-term decline were not known.

The decision analysis performed by the spectacled eider recovery team involved four stages:

Stage 1

This required the choice of a *data set*: the best available data for assessing extinction risks consisted of abundance index estimates through time. These were available only for the Yukon–Kuskokwim delta; that is, the southern Alaskan metapopulation, which forms one of the three breeding metapopulations of the species. Data were available from three independent aerial and ground surveys.

Stage 2

There must be a choice of *classification* criteria: whether a species is "endangered" or "threatened" depends on how those terms are defined, which is not determined by scientific criteria alone. Based on simulations including environmental stochasticity the spectacled eider recovery team deemed a population to be endangered provided it was declining by ≥5 % per year. The population simulations showed that a growth rate of $r = -0.05$ indicated a probability of 0.05 of reaching a critically small population of 250 adults within 50 years. The simulation used a standard deviation for r (measuring the strength of environmental stochasticity) sampled uniformly from the interval (0.07, 0.21) based on extensive data from a similar species (common eider, *Somateria mollissima*).

Stage 3

Classification must be followed by a choice of *decision* criteria: in Bayesian analysis, there must be the choice of a loss function which attributes an explicit loss to a decision; that is, to a (mis)classification. The spectacled eider recovery team considered as unprotected those populations declining at $\geq 5\%$ per year without being classified as threatened or endangered; and overprotected those populations that were maintaining a 0% or greater growth rate per year while being classified as endangered. They attributed equal probability to the two types of error. Two types of loss function were used: (1) a simple loss function which assigned a loss of one to an incorrect decision not to classify the spectacled eider as endangered and a zero loss to other decisions; and (2) a more complex continuous loss function that quantified the risk of misclassification as a function of the rate of decline.

Stage 4

Data analysis began with a Bayesian trend analysis based on an exponential growth model (Wade 1994). The use of Bayesian methods allowed the use of data from all three surveys and the explicit incorporation of the uncertainty of parameter estimates. The prior probability distribution of the population size was assumed to be a normal distribution. A uniform prior probability distribution was used for all the other parameters to reflect no prior knowledge about their values. (Thus, parameter estimation was also done using Bayes' rule.) The region for which the annual population growth rate was ≤ -0.05 encompassed 99.86% of the posterior distribution. The calculation of the expected Bayesian loss for both loss functions resulted in the population being classified as endangered in all but one analysis (with the data sets taken singly or in all possible combinations).

Taylor *et al.* (1996) compared these results with those from a decision analysis based on conventional statistics. Qualitatively, the recommendations were the same, though for the reasons mentioned at the beginning of this section, conventional statistics does not permit definite probabilities to be assigned to hypotheses.

6.5 Summary

Persistence is a key goal of systematic conservation planning. While it may be impossible to take account of all factors affecting persistence over long timescales, we can try to minimize the potential for imminent loss over shorter timescales, typically 50 years or a little more. Biological processes such as dispersal, recruitment and succession should be facilitated, habitat modification should be minimized and opportunities for populations to realize their evolutionary potential should be promoted through conservation area networks by setting appropriate targets and providing appropriate management actions. Population viability analysis, and the more recent habitat viability analysis, are two methods for addressing the likelihood of persistence of populations. These methods are data demanding and it is often not feasible to carry them out. Risk assessment and decision analysis provide less detailed methods of assessing the prognosis for biodiversity surrogates in conservation areas.

Setting good targets for representation that include persistence factors is crucial. While this is difficult, it need not be entirely arbitrary. Local experience and local ecological knowledge can help formulate rules for setting targets, as can generic reasoning about the likelihood of viability under different demographic and environmental regimes. Inevitably, decisions have to be made with high levels of uncertainty. A formal decision analysis technique – Bayesian decision analysis – allows the quantitative incorporation of uncertainty during parameter estimation, model evaluation and the decision-making context itself. Though it has its critics, Bayesian decision analysis provides a rational framework for making decisions when there are many options to be considered.

7 · *Satisfying multiple criteria*

Biodiversity conservation is not the only potential use of land. Other potential uses include recreation, human habitation, habitat transformation for agricultural or industrial development, biological and industrial resource extraction, etc. (Sarkar 2004). Some of these other uses are compatible with biodiversity conservation, for example, recreation may consist of viewing biodiversity unobtrusively. Others, for instance the conversion of habitat into croplands, industrial parks or strip-mining, are clearly not. In many contexts, it would be unethical for conservation-area network design to ignore these other criteria; in most contexts it would at least be politically imprudent. Returning to an example discussed at the beginning of Chapter 5 (Section 5.1), the designation of the avifauna-rich wetland Keoladeo Ghana in India as a national park in 1981 resulted in the abrogation of traditional pasture rights of impoverished local communities (Lewis 2003). There is ample ethical reason for rejecting such an action particularly because, in this case, ecological studies had not been carried out first to test whether the exclusion of grazing was even necessary for better management of the wetland as bird habitat. In fact, as noted earlier, subsequent ecological studies showed that grazing was in fact necessary to maintain bird species diversity (Vijayan 1987). Meanwhile, during protests over the designation of the park, nine people were killed by the Indian police. Even leaving the obvious ethical considerations aside, in cases such as this, it is imprudent to ignore competing claims for land: local opposition can undermine even the best-intentioned conservation plans. (For more details on such ethical questions see Norton (1987) and Sarkar (2005).)

This chapter will describe some of the best-known methods for attempting to optimize simultaneously biodiversity representation and these other competing objectives for the use of land. Besides negotiating these socio-political objectives, these same methods can also be used to incorporate biological design criteria for conservation area networks that go beyond representation, for instance spatial configuration criteria related to persistence such as the size and shape of individual conservation areas, their dispersion across the landscape and their connectivity. Other vulnerability and threat criteria such as environmental impacts and accessibility can also similarly be incorporated.

Typical socio-political objectives include the minimization of economic losses and the social costs of implementing a conservation plan. Economic costs include the costs of acquiring areas and the costs of implementing conservation-management plans on areas as well as the forgone opportunity in not exploiting the resources of an area in order to protect its biodiversity (Faith 1995; Sarkar 2004). The first two of these are easy to estimate accurately but estimation of the third is more difficult and typically requires recourse to economic models. Social costs include the impact of a plan on people, which may be at least crudely estimated by the number of people affected by the plan. However, there are socio-political objectives that might be important in conservation-area planning, which cannot be easily interpreted in terms of such costs. These objectives include the maximization of scenic value, wilderness value, etc.

Among the biological criteria, some spatial design criteria are easily quantified. For example, size can be estimated by the area of individual conservation areas or entire networks. Shape is less straightforward. However, if compact shapes are to be preferred, it may be estimated reasonably accurately by the ratio of boundary length to area. Connectivity and dispersion are more difficult to quantify and estimate though not impossible – for further discussion, see Forman (1995). All the vulnerability criteria discussed in the last chapter, including environmental impact and accessibility, can be quantified and used in this context.

Thus the objectives other than biodiversity representation that may be relevant to the design of conservation plans can be many and varied. Over the last three decades the decision-theory community has devised a wide variety of methods for the incorporation of multiple criteria – the field has variously been called multi-criterion (or multi-criteria) decision-making (MCDM) (Arrow & Raynaud 1986), or multiple objective (or multi-objective) decision-making (Jannsen 1992; Keeney & Raiffa 1993; Bogetoft & Pruzan 1997). Here, Arrow and Raynaud's original terminology will generally be used, though the process will be called multi-criteria analysis (MCA). Available methods range from heuristic multi-dimensional optimization algorithms to the well-developed multi-attribute value and utility theories (MAVT and MAUT) Dyer *et al.* 1992; (Keeney and Raiffa 1993; Dyer 2005). We will not survey all available methods but restrict our attention to those which are commonly used in conservation planning. (For a review of the use of MCA techniques in conservation-area network design, see Moffett and Sarkar (2005).)

7.1 Iterative- and terminal-stage procedures

Typically, though not always, conservation area networks are constructed iteratively. This means that individual areas are selected one or a few at a time for inclusion in a potential network of areas. For instance, complementarity-based algorithms select areas in this way. Consequently, there can be two types of protocol for the incorporation of multiple criteria into systematic conservation planning: (1) in *iterative*-stage protocols all criteria are considered as each individual area is selected for potential inclusion (Faith *et al.* 1996; Faith & Walker 1996b); and (2) in *terminal*-stage protocols a collection of potential conservation area networks is initially selected, with each potential network satisfying the biodiversity representation targets (Sarkar *et al.* 2004a). Criteria other than biodiversity representation are then use to rank these potential networks. In iterative-stage protocols, each area that may be included in a network comprises a "feasible alternative" or, in short, "alternative." (In this context, the term "feasible" indicates that an alternative should not be precluded from further consideration.) In terminal-stage protocols, an entire potential network of areas that satisfies all biodiversity representation criteria constitutes such an alternative.

Terminal-stage protocols typically (though not always) give biodiversity representation primacy over other criteria by requiring that the satisfaction of the biodiversity representation targets cannot be compromised. Iterative-stage protocols may not give biodiversity representation primacy. They may allow trade-offs to be made between biodiversity representation and other criteria. The two protocols are not mutually exclusive. Some criteria can be incorporated at the iterative stage through trade-offs while others can be incorporated at the end. Whether an iterative-stage or terminal-stage protocol is appropriate depends on a partly subjective decision that can involve many factors, for instance the perceived relative importance of the biodiversity surrogates being used, and the targets set for them, compared to all the other criteria. If these surrogates and targets are viewed as extremely important, then we may prefer a terminal-stage protocol which does not allow any compromise on the satisfaction of biodiversity representation goals, but this may be at the expense of achieving a compromise that is more likely to be implemented.

7.2 The valuation framework

The framework for multi-criteria analysis (MCA) consists of a goal, a set of feasible alternatives, $A = \{\alpha_j : j = 1, 2, \ldots, m\}$ and a set of criteria, $K = \{\kappa_i : i = 1, 2, \ldots, n\}$. The goal will be context-dependent. It may be the selection of a single "best" alternative, it may be the selection of a subset of alternatives with or without an internal ranking or it may be a ranking of the entire alternative set, etc. The goal need not be formally described since it does not formally enter into the description of MCA methods, though it must be clearly stated for a planning process to make sense. Moreover, since MCA methods are based on an abstract characterization of the criteria, in what follows, we will not distinguish between biological and socio-political criteria. However, Table 7.1 lists these criteria separately and provides a list of the criteria that have been used for conservation-area network design in the literature.

Two criteria are *compatible* if they rank all the alternatives in the same order, otherwise they are *incompatible* (Sarkar 2005). If all criteria are compatible then they would rank all alternatives in the same order and there would thus be no problem of synchronizing multiple critera. Typically, most of the criteria we consider, for instance, biodiversity representation, economic cost and social cost are not compatible with each other. A criterion is called *commensurable* if alternatives can be placed in a linear order using that criterion. We will restrict attention to commensurable criteria because non-commensurable criteria cannot be used in a systematic procedure. Many social and personal values are considered to be incommensurable though economists often try to measure them by eliciting preferences from individuals through methods such as contingent valuation (Keeney & Raiffa 1993; Sarkar 2005).

We must assume that every alternative can at least be ordinally (or qualitatively) ranked by each criterion; that is, given any two alternatives, α_e and α_f, and a criterion, κ_l, α_e is either better than, as good as, or worse than α_f by κ_l. More formally, every criterion introduces a linear (or complete) weak order on A. (The order is *weak* because α_e can be as good as α_f; in a *strict* order, α_e would be either strictly better than or strictly worse than α_f.) This requirement is reasonable because a criterion that does not at least order all the alternatives cannot be used to assess all of them. This minimal assumption is sufficient for the development of what turns out to be an important MCA method, non-dominated set (NDS) computation, will be described in the next section (Sarkar *et al.* 2000; Sarkar & Garson 2004).

Table 7.1 *Criteria used in conservation-area planning*

Biological criteria	Socio-political criteria
Biodiversity surrogate representation[a]	Economic cost[b]
Size of individual units[c]	Recreational value[d]
Total area[e]	Human population[f]
Shape[g]	Future economic value[h]
Dispersion[i]	Scenic beauty[j]
Connectivity[k]	Cultural heritage[l]
Environmental impact[m]	Educational value[n]
Accessibility[o]	

[a] No additional references are included because every CAN design exercise in this table includes the representation of biodiversity.
[b] Mendoza and Sprouse (1989); Laukkanen *et al.* (2002).
[c] Sarkar *et al.* (2004); Moffett & Sarkar (2005).
[d] Mendoza and Sprouse (1989); Malczewski *et al.* (1997); Li *et al.* (1999); Huang *et al.* (2002); Villa *et al.* (2002); Ananda and Herath (2003); Herath (2004); Redpath *et al.* (2004); Janssen *et al.* (2005).
[e] Rothley (1999); Noss *et al.* (2002); Sarkar *et al.* (2004b); Moffett *et al.* (2006).
[f] Malczewski *et al.* (1997); Sarkar *et al.* (2000); Faith *et al.* (2001a); Sarkar and Garson (2004).
[g] Noss *et al.* (2002); Sarkar *et al.* (2004b); Moffett *et al.* (2006).
[h] Kuusipalo and Kangas (1994); Berbel and Zamora (1995); Faith *et al.* (1996); Malczewski *et al.* (1997); Li *et al.* (1999); Faith *et al.* (2001a); Huang *et al.* (2002); Laukkanen *et al.* (2002); Villa *et al.* (2002); Ananda and Herath (2003); Herath (2004); Huth *et al.* (2004); Redpath *et al.* (2004); Janssen *et al.* (2005).
[i] Sarkar *et al.* (2004b).
[j] Laukkanen *et al.* (2002); Bojórquez-Tapia *et al.* (2004); Redpath *et al.* (2004).
[k] Rothley (1999); Geneletti (2004); Sarkar *et al.* (2004b).
[l] Redpath *et al.* (2004); Janssen *et al.* (2005).
[m] Mendoza and Sprouse (1989); Malczewski *et al.* (1997); Li *et al.* (1999); Huang *et al.* (2002); Janssen *et al.* (2005); Phua and Minowa (2005).
[n] Li *et al.* (1999); Bojórquez-Tapia *et al.* (2004); Redpath *et al.* (2004).
[o] Malczewski *et al.* (1997); Li *et al.* (1999); Sierra *et al.* (2002); Villa *et al.* (2002); Bojórquez-Tapia *et al.* (2004); Geneletti (2004); Redpath *et al.* (2004); Sarkar *et al.* (2004b); Moffett *et al.* (2006).

Beyond qualitative ordering, the next question is whether each alternative can be assigned a quantitative value by each criterion. As Figure 7.1 shows, most existing MCA methods assume that this can be done. In general there are two sources of uncertainty that may prevent attributing quantitative values to alternatives: (1) imprecise definition of alternatives

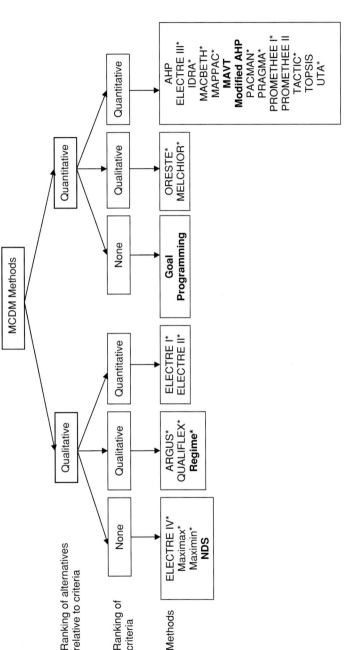

Figure 7.1. A taxonomy of MCDM methods based on requirements placed upon criteria and alternatives. To keep the figure simple, we have eschewed repetition. Throughout, whenever there is a branch, all hanging lists on the left can also be included in the corresponding lists to its right. Thus, it is possible to use NDS computation when there is a qualitative or quantitative ranking of preferences and when the alternatives can also be quantitatively ranked, it is possible to use Regime when there is a quantitative ranking of preferences and when the alternatives can also be quantitatively ranked, and so on. A "*" indicates that this method has never been used in conservation planning as far as we know. Methods indicated in bold are recommended by Moffett and Sarkar (2005). (From Moffett & Sarkar (2005). Reproduced with permission from Blackwell Publishing.)

and (2) imprecise measurability of alternatives. In the context of conservation planning the former problem is rarely relevant. Alternatives, either as individual areas or as entire potential conservation-area networks, are well defined. However, the question of measurability is relevant. Some socio-political criteria such as economic cost of a plan and some spatial design criteria such as the size of selected units are relatively easy to quantify. Others, including socio-political criteria such as social cost or even the opportunity costs of a plan, as well as spatial design criteria such as shape or connectivity, are not so easily quantifiable without adopting potentially debatable conventions. Consequently, we often have more confidence in the (qualitative) order of our alternative rankings than in quantitative values. This is what generates interest in the left branch of the tree in Figure 7.1 even though, the moment we either elicit quantitative preferences or otherwise approximate qualitative rankings by quantitative values (sometimes called "pseudo-metric" values) many computationally simpler MCA methods become available.

Independent of ranking alternative networks of areas or alternative areas within networks, is the question of ranking the criteria themselves. Now we must consider all three possibilities: that there is no ranking, that there is a qualitative ordinal ranking and that the criteria may be assigned quantitative values. In the case of criterion rankings, quantitative values are typically even more difficult to assign than in the case of ranking alternatives. Assume that the alternatives are only qualitatively ordered. Figure 7.1 shows the major MCA methods available for use in conservation planning. If the criteria cannot be ordered at all, the methods available include Dominance or NDS computation (see Section 7.3). If the criteria can only be qualitatively ordered, the Regime method (Hinloopen *et al.* 1983), which also does not make ad hoc assumptions, is available as an option. All other methods on the left branch of Figure 7.1 have to make at least some arbitrary assumptions about alternative or criterion ranks (Moffett & Sarkar 2005).

If it is assumed that the alternatives can be quantitatively assessed, all the methods that have so far been mentioned can still be used (because quantitative ordering logically entails qualitative ordering). Beyond these, there exist very few methods that do not also require quantitative ordering of the criteria. Section 7.4 will discuss methods based on these assumptions. Assuming that all such quantitative assignments are possible, and that these assignments are rational, amounts to a commitment to the standard economic model of estimating preferences. It allows the co-option of all the techniques that have been developed for the elicitation of preferences, etc., but is also subject to the standard objections to the economic model of rationality. For a good discussion of attendant problems, see Bateman and Willis (1999). Figure 7.2 shows how our knowledge in a given decision-making context can be used to select an appropriate MCA method.

7.3 Non-dominated alternatives

The simplest MCDM method, which is nevertheless fully adequate for many contexts, is Dominance or the non-dominated set (NDS) computation for the alternatives. One alternative "dominates" another alternative if it performs better with respect to at least one criterion and performs at least as well with respect to all other criteria. The set of "non-dominated" alternatives consists of those alternatives that are not dominated by any other alternative. Thus, this set consists of the unequivocally superior alternatives: regardless of the particular importance attributed to the each of the individual criteria, an alternative that is dominated will never outperform all non-dominated alternatives (Sarkar & Garson 2004). Therefore, in selecting a network of conservation areas, the alternatives that merit further consideration should be limited to this set of non-dominated alternatives. Such a use of the non-dominated alternative set in conservation planning goes back at least to Rothley (1999) and was explicitly advocated by Sarkar *et al.* (2000) and Sarkar and Garson (2004).

The NDS corresponds to the Pareto optimal sets of traditional economic analysis (Keeney & Raiffa 1993). The evaluation of a set of alternatives through the calculation of an NDS has the following advantages: (1) it requires only that each criterion induce a weak linear ordering on the alternatives, and thus does not require (a) the assignment of quantitative values to the alternatives, (b) an evaluation of the relative importance of the criteria, or (c) an independence assumption requiring that the preferences for values on one criterion are not influenced by values on the other criteria; (2) it introduces no subjective information into the decision-making process other than that required to produce the weak orderings of the alternatives; and (3) its results are compatible with those of any other rational decision procedure. As the examples below will show, the NDS can be easily computed using a variety of software packages.

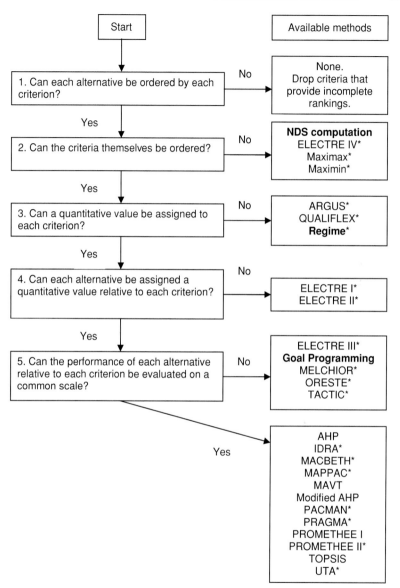

Figure 7.2. A decision procedure for the selection of an existing MCDM method. A "*" indicates that this method has never been used in conservation planning as far as we know. Methods indicated in bold are recommended by Moffett and Sarkar (2005). (From Moffett and Sarkar (2005). Reproduced with permission from Blackwell Publishing.)

Example 7.1 Connectivity, area and representation in Nova Scotia

Rothley (1999) used NDS computation with the goal of identifying five forest nature reserves from a set of 20 potential reserves to supplement two existing national parks in Nova Scotia, Canada. Three criteria were used first in a terminal stage protocol: (1) connectivity; (2) total area; and (3) representation of rare plant species. The goal was to maximize all three criteria in the finally selected set. There are 15 504 possible combinations of five reserves from a set of 20. Numerical values were assigned to each of the combinations or alternatives on the basis of each of the criteria. Connectivity was measured by the inverse of the distance between two included potential reserves. Area measurement was straightforward. Representation was measured by the number of rare species present in the five potential reserves. An optimization algorithm (multi-objective programming [Cohon 1978]) was used to reduce this set of 15 504 alternatives to 36 non-dominated alternatives. This part of the analysis was performed using the Lindo/386 5.3 (Lindo Systems 1995) software package. Clearly in this case the NDS is probably too large for use in the next stage of planning and should be further refined. Rothley (1999) goes on to attempt such a refinement with a trade-off analysis similar, in principle, to the one discussed in Section 7.4.

Example 7.2 Social and economic costs in Texas

A second example used the modeled spatial distributions of 655 animal species in Texas produced by the Texas GAP Analysis Project and analyzed by Sarkar et al. (2000). These distributions were known for 1183 hexagons, each with an average area of 649 km². The goal of this planning exercise was to develop a method for minimizing social and economic costs while ensuring adequate biodiversity representation. A terminal-stage protocol was used with the 655 animal species as surrogates for biodiversity. One hundred different plans or alternatives were generated each using a representation target of at least 10 % of the distribution patterns of each species. These computations were done using the ResNet software package (Garson et al. 2002b). For each such alternative, two additional criteria were used: (1) the projected human population in 2020 in that set of hexagons and (2) the total area of the set. Criterion (1) was interpreted as a measure of the social

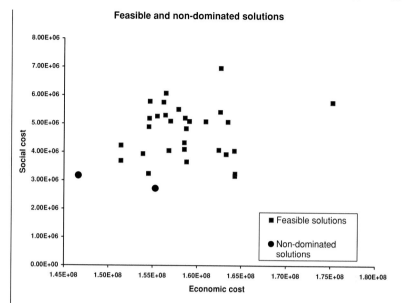

Figure 7.3. Non-dominated set (NDS) computation. When only two criteria are involved, the NDS can be identified by visual inspection. For more details see the text. (Redrawn using data from Sarkar *et al.* (2000).)

cost of conservation, which should be minimized. (Alternatively, it may be interpreted as a measure of future threat.) Criterion (2) was taken as a measure of the economic cost of conservation, which should also be minimized. There were two non-dominated solutions, which are shown in the MCS plot of Figure 7.3. This part of the analysis was performed using the MultCSync software package (Sarkar *et al.* 2004a; Moffett *et al.* 2006). Such MCS plots are useful visual devices when there are only two criteria. If the alternative values are being minimized for both criteria, the non-dominated alternatives are those with no others in the lower left-hand quadrant below them. (Similar arguments can be made for the other three quadrants if one or both of the criteria are being maximized.) However, this visual method cannot be extended to more than these dimensions.

If the number of non-dominated alternatives is small, the non-dominated alternative set can be presented to political decision-makers who can then select among them on the basis of other considerations (Sarkar & Garson 2004). However, in many decision scenarios, the number of non-dominated alternatives will be too high for this purpose.

In such scenarios, the identification of the preferred alternative from this set will require a more refined evaluation of the alternatives than that afforded by just calculating the non-dominated alternative set.

7.4 Refining non-dominated sets

As just noted, if the number of non-dominated alternatives is small, they can all be presented to decision-makers with no further analysis. However, typically, the number of non-dominated alternatives increases rapidly with the number of criteria (Sarkar & Garson 2004). Thus, the number of non-dominated alternatives may be intractably large for use during the decision-making process. It then becomes imperative to rank the non-dominated alternatives so that some of them can be eliminated. We will restrict attention to the two most common methods. Both involve the construction of an "objective" function measuring the value of alternatives. The goal of the methods is to maximize the value of this function over the alternative set.

7.4.1 Trade-off analysis

In trade-off analysis weights are assigned to each criterion, alternatives evaluated using these weights and then the weights are varied to see how the performance of alternatives changes. In a typical application in our context the biodiversity value of each alternative (as measured by complementarity) is compared to the weighted value of the area measured by the other criteria (for example, economic cost). Areas are justified for inclusion if the biodiversity value exceeds the weighted cost according to these other criteria. The weights on the other criteria can be iteratively lowered to see how much more biodiversity value can be achieved at a higher cost with respect to the other criteria. Example 7.3 shows how a trade-off analysis is carried out.

7.4.2 The modified analytic hierarchy process (mAHP)

The analytic hierarchy process (AHP) (Saaty 1980) is a decision procedure used to rank order a set of alternatives on the basis of multiple criteria. The AHP has found extensive use in conservation planning (see below). To use the AHP, a quantitative value must be assigned to each alternative relative to each criterion, and the importance of each criterion must be quantitatively assessed. In order for the results of the AHP to be meaningful, the criteria themselves must exhibit mutual difference

Example 7.3   Trade-off analysis for coastal eucalyptus forest in south-eastern Australia (Plate 2)

Faith *et al.* (1996) used forest production opportunities in south-eastern Australia as costs and compared these costs with biodiversity values for areas calculated using complementarity. Potential conservation (or forestry) areas were 2914 0.01 × 0.01 degree grid cells. They summed a number of variables referring to suitability for forestry from across the entire forestry value chain, such as site suitability, tree species, potential for regeneration and distance to mill. They measured biodiversity with environmental surrogates by calculating the similarity or difference (association; see Chapter 4) between each grid cell and every other grid cell based on 17 temperature, precipitation, solar radiation, nutrient index and terrain roughness variables. They ran a principal coordinate analysis on the association matrix to derive a reduced dimensional environmental space (Gower 1966) and located each grid cell in this ordination space.

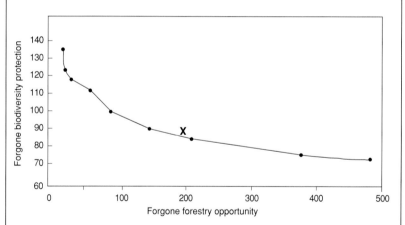

Figure 7.4. A trade-off curve in which the points from left to right represent solutions for weights, *w*, on forestry of 1.0, 0.8, 0.7, 0.5, 0.25, 0.1, 0.06, 0.02 and 0.01. For each area, a forestry suitability value was multiplied by *w* to convert it to effective-biodiversity units. The search for solutions used a simple heuristic; results for 1.0 and 0.5 are not global optima. The numbers of areas chosen for protection for each solution-point (from left to right) are 4, 5, 6, 8, 12, 22, 31, 54 and 69. If the weight was exactly 0.0, reflecting no value on forestry, all areas would be selected for protection. The "*x*" represents a solution found for 22 protected areas, ignoring costs. (Redrawn from Faith *et al.* (1996). Reproduced with permission from Elsevier.)

They ranked alternatives from high opportunity cost and high bio-diversity protection, through a series of trade-offs representing alternatives with lower cost and lower biodiversity protection through to lowest cost and very low biodiversity protection. Areas were selected for protection if their complementarity valued exceeded the weighted cost. The process was repeated for a range of different weights on cost. Figure 7.4 shows the resulting trade-offs curve. For any given cost, the points along this curve are as close as it is possible to get to both low forgone forestry opportunities and low forgone biodiversity. The "**x**" in Figure 7.4 represents a solution found ignoring costs. This solution required 22 areas. However, the point on the trade-offs curve for 22 areas is a better solution in that it represents less forgone (i.e., more) forestry opportunity and about the same forgone biodiversity.

independence (Dyer 2005). (This means that the preference difference between two alternatives differing with respect to some subset of criteria does not depend on the shared values of the alternatives on the other criteria, and justifies the use of an additive model of preference.)

In standard multi-attribute value theory (MAVT) or multi-attribute utility theory (MAUT) a utility function is constructed to rank the non-dominated alternatives on the basis of their utility values (Keeney & Raiffa 1993; Dyer 2005). The AHP avoids the explicit construction of such a function. Instead, it elicits values on the users' implicit preference function by requiring a numerical pairwise comparison of the criteria on an increasing ratio scale, usually from 1 to 9 (Saaty 1980). This approach then generates weights, or scaling constants, for the criteria using the pairwise binary comparisons. A value of one indicates that the two criteria being compared have the same rank; a value of nine indicates that changes over the range of values for the second are maximally preferred to changes over the range of values for the first. Thus, if criterion (A) has a ratio scale value of X compared to criterion (B), then criterion (B) has a ratio scale value of $1/X$ compared to criterion (A). These comparisons of the importance of each criterion should be based on the following question: how much better would it be to improve the performance on one criterion from its worst to its best value (as defined by the ranges of the criterion values among the alternatives under consideration) versus improving the performance on another criterion from its worst to its best value? Note that the notion of the importance of a criterion has no meaning without reference to the range over which it varies (Keeney & Raiffa 2003). The results of these pairwise comparisons are used to assign a quantitative value, ω_i, to each criterion i. The results can then be represented as an $(i \times i)$ matrix. The components of the normalized eigenvector of this matrix with the highest eigenvalue provide the rankings of the criteria. (This is essentially one approach to averaging the redundant comparisons.)

Let v_{ij} be the value of alternative α_j relative to criterion κ_i. Each v_{ij} also may be obtained from pair-wise comparisons of the performance of each alternative on each criterion. Each v_{ij} is normalized, yielding v'_{ij}. The value of each alternative, π_j, is then calculated as follows:

$$\pi_j = \sum_{i=1}^{n} \omega_i v'_{ij} \qquad (7.1)$$

(where ω_i is the weight of the i-th criterion).

Depending on the procedure used to normalize the v_{ij}, it is possible to reverse the rank order produced by the AHP through the addition of a new alternative to the set of alternatives (Belton & Gear 1983). However, rank reversal can be prevented by normalizing the values of the alternatives according to each criterion in a way that does not depend on the number of alternatives. The normalization formula given here was proposed by Dyer (1990) (which is similar to a procedure proposed by Kamenetzky 1982), whereby each v_{ij} is normalized to

$$v'_{ij} = \frac{v_{ij} - \min[v_{ij}]}{\max[v_{ij}] - \min[v_{ij}]} \qquad (7.2)$$

where "$\min[v_{ij}]$" and "$\max[v_{ij}]$" are the minimum and maximum possible values of v_{ij} for criterion κ_i, and they define the ranges of the criterion values. This procedure has been used in conservation planning by Pereira and Duckstein (1993), Phua and Minowa (2005) and Moffett *et al.* (2006). Example 7.4 shows its use.

In conservation planning, as in many other disciplines, the original AHP has been a very popular technique for assessing these weights and for the calculation of these criterion values (Anselin *et al.* 1989; Kangas 1993; Li *et al.* 1999; Mendoza & Prabhu 2000; Schmoldt *et al.* 2001; Clevenger *et al.* 2002; Villa *et al.* 2002; Ananda and Herath 2003). However, due to its susceptibility to rank reversal, the original AHP is flawed. The fact that the addition of a new alternative to the set of those under consideration can alter their ranking makes the AHP an arbitrary method of decision-making. This fact has been known by decision-theory researchers for some time and the fact that the AHP is used with such regularity in the context of conservation planning is at least somewhat surprising given the existence of comparable methods that provide the ease of use associated with the AHP while nonetheless avoiding the possibility of rank reversal (Dyer 1990). The modified AHP (mAHP) technique described above is one such method. An added benefit of this method is that it produces results that are consistent with results produced by MAVT and MAUT, methods that are well accepted by economists and decision analysts (e. g., von Winterfeldt & Smith 2004).

Example 7.4   A conservation plan for Ecuador (Plates 11 and 12)

Sarkar *et al.* (2004a) developed a conservation plan for continental Ecuador (excluding the Galápagos Islands) using the mAHP. With an area of 248 750 km^2, this region is small in size but rich in biodiversity (Sierra *et al.* 2002). Since geographical distributions of species were not available for a representative set of taxa, planning was based on modeled distributions of coarse biological surrogates. This analysis started with a 200 × 200 m^2 raster grid on which the modeled spatial distributions of 46 major vegetation types were mapped. These vegetation types span the entire floral range of Ecuador. Sierra (1999) and Sierra *et al.* (2002) provide details on the classification and modeling of the distribution of the vegetation types. At this spatial scale, each data cell contained one vegetation type. This scale of resolution was reduced to a 2 × 2 km^2 grid in which each new cell consisted of 100 of the original cells. There was a total of 37 727 cells. The motivations for the scale change were to improve computational efficiency because of the reduced size of the data set and to use areas that are of appropriate size to be regarded as units of conservation.

The analysis kept track of the vegetation types in each of the original cells that were compounded to make a new cell. Each of the new cells could potentially contain at most 46 vegetation types. For each cell, for each vegetation type, the probabilistic expectation of the presence of that type in that cell was set equal to its proportion in the original 100 cells. Thus, if all the original cells contain exactly the same vegetation type, then that type has an expectation equal to one and each other type has an expectation of zero.

The map of Ecuador was further modified by masking areas that were permanently transformed by anthropogenic modification as of 1996 (Sierra *et al.* 2002) and were, therefore, inappropriate for inclusion in a conservation plan. In this way 39 % of the cells were excluded. The Ecuadorian national reserve system (NRS) was also represented on a

$2 \times 2\,\mathrm{km}^2$ grid. The target of representation for each vegetation type was set to 10 % of the untransformed area in which that type occurred. Thirteen of the 46 vegetation types did not meet this target within the NRS.

Place prioritization

Places were prioritized using the ResNet software package (Garson *et al.* 2002b) with areas being selected using rarity, complementarity and adjacency rules used hierarchically. This preference for adjacency results in larger conservation areas. The selection procedure was initiated using the existing NRS of Ecuador. Thus, the final solution records the minimum number of cells that must be added for the satisfaction of those targets according to this heuristic algorithm. One hundred different solutions were generated using randomized re-orderings of the data set. These re-orderings generated different solutions because they resulted in the selection of different cells when ties were broken by lexical order (that is, by selecting the next cell in the list of cells). Figure 7.5 shows two of the solutions generated in this fashion.

Multiple criteria

The second stage of the analysis consisted of incorporating other criteria to rank the feasible alternatives. There were three steps: (1) an identification of the relevant criteria and the ranking of the alternatives according to each criterion; (2) the determination of the NDS; and (3) further refinement of this set to find a final set of preferred alternatives. Six criteria were used:

(1) *The aggregate number of conservation areas*, which should be minimized to achieve spatial cohesiveness of conservation area networks;
(2) *The average area of each conservation area*, which should be maximized to encourage larger conservation areas. (This aspect of conservation planning was also encouraged by the use of the heuristic rule preferring adjacency earlier);
(3) *The variance of the areas*, which should be minimized to further discourage the selection of very small areas;
(4) *The aggregate distance of the selected cells to existing components of the NRS*, which should be minimized, again to increase cohesiveness (the distances being calculated between the centroids of the nearest cells);

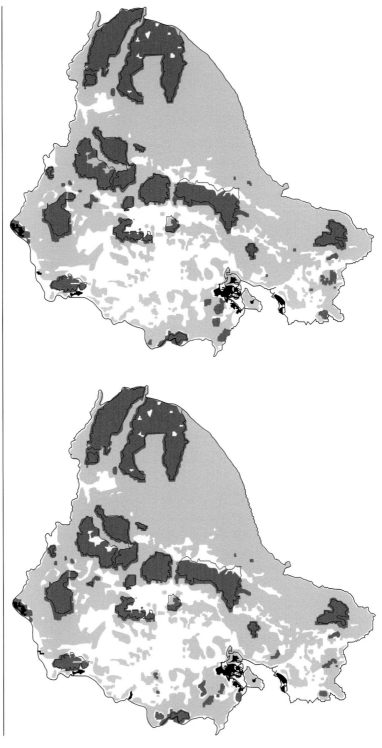

Figure 7.5. Potential conservation area networks for Ecuador. These are the two non-dominated solutions that receive the highest ranking by the modified analytic hierarchy process. For further detail see the text. (Redrawn using data from Sarkar *et al.* (2004a). Reproduced with permission from University of Michigan – School of Nature Resources & Environment.)

(5) *The aggregate distance to anthropologically transformed areas*, which should be maximized to decrease the threat of habitat destruction (the distances once again being calculated from the centroids of the nearest cells);
(6) *The total area of the selection cells*, which should be minimized to decrease the cost of acquisition of the added cells.

Criteria (1)–(4) are biological with the first four evaluating aspects of spatial configuration criteria; criterion (5) is a persistence criterion. Criterion (6) is socio-political. All 100 feasible alternatives were evaluated according to each of these criteria, which are such that a definite quantitative (numerical) value could be assigned to each alternative. There were 58 non-dominated alternatives, clearly too many to be handed to political decision-makers in most contexts. (The alternatives shown in Figure 7.5 are both non-dominated). In Step 3 each alternative was numerically ranked according to each criterion, and the criteria themselves must be numerically ranked. The ratio scale ranking of the six criteria, taken in order, was represented by the following matrix:

$$
\begin{pmatrix}
1 & 1/2 & 9 & 3 & 6 & 7 \\
2 & 1 & 9 & 9 & 4 & 9 \\
1/9 & 1/9 & 1 & 1/5 & 1/6 & 1/2 \\
1/3 & 1/9 & 5 & 1 & 1/3 & 4 \\
1/6 & 1/4 & 6 & 3 & 1 & 2 \\
1/7 & 1/9 & 2 & 1/4 & 1/2 & 1
\end{pmatrix}
$$

The eigenvector of this matrix with the highest eigenvalue provides the rankings of the criteria. The rankings presented here were developed by Sarkar *et al.* on the basis of field experience. The software package MultCSync was used for this purpose. MultCSync allows a check of the consistency of the elicited rankings, so that the elicitation process can be iterated until an acceptable consistency level is found. The rankings presented in the matrix were of acceptable consistency according to the standards of the package.

The two alternatives shown in Figure 7.5 were the two best alternatives found in this way, taking all six criteria into account and using Equations (7.1) and (7.2). They select different areas in southwestern Ecuador thus potentially offering alternative choices to political decision-makers. Since all non-dominated alternatives were ranked, a set of best alternatives of any cardinality was available for presentation to such decision-makers.

7.5 Sensitivity analysis

Once we move beyond the identification of the non-dominated set, consideration of multiple criteria in the development of a conservation area network typically requires the evaluation of multiple alternatives on the basis of multiple criteria by constructing a linear additive value function using quantitative evaluations of both the criteria and the alternatives. This is the approach taken by the mAHP and other techniques based upon MAVT (Belton 1986). Methods of this type assign a quantitative value to each alternative and each criterion in order to provide a rank order of the alternatives. Such methods pay for this increased precision with an accompanying increase in subjectivity, coupled with the employment of a more complicated methodology. To mitigate the effects of subjectivity in such assignments, the use of these methods should always be accompanied by sensitivity analyses to test the robustness of the final rankings. No single methodology for sensitivity analysis accommodates all sources of uncertainty. Therefore, whenever possible, a variety of such methodologies should be used (Butler et al. 1997).

There are two standard methods of sensitivity analysis, and it remains a goal for the future to devise others precisely tailored to the selection of conservation area networks. The first is single-dimensional analysis – the effect of altering the weight assigned to a single criterion is evaluated by increasing, for each criterion, the ω_i assigned to the criterion from zero to one while holding constant the relative weights assigned to all other criteria. The trade-off between biodiversity contribution and timber volume used in the Papua New Guinea study in Chapter 8 (Section 8.5) is an example of this one-dimensional analysis. An area was admitted to the list of selected areas if the weighted timber volume (cost) was less than the complementarity value of that area. The method used to make this trade-off is described in Chapter 8 (Section 8.5).

The second is multi-dimensional simulation-based rank order analysis – the effect of altering the weights assigned to multiple criteria is evaluated by generating a large number of sets of random weights and assigning the weights in each set to the criteria in a way that preserves the rank ordering of the assessed weights. For example, the greatest weight in the random set may be assigned to the criterion with the greatest assessed weight, while the smallest weight in the set may be assigned to the criterion with the smallest assessed weight. The new rank orderings produced in this way are then compared with the original rankings. Example 7.5 demonstrates how this can be done. For a detailed discussion of the use of simulation to test the robustness of a ranking based on multiple criteria and other multi-dimensional methods, see Butler et al. (1997).

Example 7.5   North-central Namibia (Plates 9 and 10)

Moffett *et al.* (2006) used data from north-central Namibia produced by the Ministry of Lands, Resettlement and Rehabilitation of Namibia. The region consisted of Etosha National Park and the land between it and the border with Angola (see Figure 7.6). The goal was to generate a plan optimizing six criteria (see below) while representing 10% of the habitat of each vegetation type in selected areas including the Etosha Park. Besides Etosha there were 119 different cells under consideration. These cells varied in size from 0.02 km² to 1225.89 km², with an average area of 517.95 km². All 35 different vegetation classes from this region were used as biodiversity surrogates and probabilistic expectations of presence were calculated for each of the surrogates in each of these 120 different cells, with the expectation value for a given vegetation class in a cell being equal to the area in the cell containing the vegetation class.

Besides biodiversity representation, they used six other criteria in a terminal stage protocol: (1) area; (2) human population; (3) number of summer cattle; (4) number of winter cattle; (5) farming; and (6) number of wildlife. An optimal solution was supposed to minimize the values of criteria (1) through (5) while maximizing the value of criterion (6). Each alternative was assigned a quantitative value for each of the six criteria using data from the website mentioned above. This set of criteria was deemed to be the most relevant by local Namibian experts.

Place prioritization, the NDS and the AHP

The ResNet software package (Garson *et al.* 2002b) was used to generate 94 different solutions or alternatives as in Example 7.4. Similarly the NDS was computed using the MultCSync software package (Sarkar *et al.* 2004a; Moffett *et al.* 2006) and consisted of 49 alternatives. Personnel from the Namibian Ministry of Lands, Resettlement and Rehabilitation provided the pairwise comparisons of the criteria which were used to assign weights to the criteria, as shown in Table 7.2, for the use of the modified AHP. The highest ranked alternative is shown in Figure 7.6a. Relatively little separated the set of cells associated with this alternative from the sets associated with the other top alternatives. Of the five highest ranked alternatives, only one differed from this one by more than one cell. It was ranked fourth and differed from the highest alternative by two cells. It is shown in Figure 7.6b.

Table 7.2 *Calculation of criteria weights. The calculation of weights for each of the six criteria used to evaluate conservation area networks in north-central Namibia is shown below. The matrix was constructed by comparing the importance of each criterion with that of every other criterion on a scale of $1/9$ to 9 by an individual familiar with the ranges over which these criteria varied. Entry (i, j) then represents the ratio by which criterion κ_i is evaluated to be more important than criterion κ_j. For example, criterion 1 (area) is evaluated to be half as important as criterion 2 (the human population) after considering these ranges; thus, the entry (1, 2) = $1/2$. The adjoining vector is the eigenvector with the highest eigenvalue of the matrix; eigenvector components are normalized such that they sum to one. The weight of the i-th criterion, ω_i, is set equal to the i-th entry of the eigenvector*[a]

$$
\begin{pmatrix}
1 & 1/5 & 1 & 1/3 & 1/4 & 1/8 \\
5 & 1 & 1 & 1 & 4 & 1/2 \\
1 & 1 & 1 & 1 & 1/3 & 1/7 \\
3 & 1 & 1 & 1 & 1 & 1/3 \\
4 & 1/4 & 3 & 1 & 1 & 1/2 \\
8 & 2 & 7 & 3 & 2 & 1
\end{pmatrix}
\Rightarrow
\begin{pmatrix}
0.047927 \\
0.196843 \\
0.080739 \\
0.134108 \\
0.143483 \\
0.396900
\end{pmatrix}
\Rightarrow
\begin{matrix}
\omega_1 = 0.047927 \\
\omega_2 = 0.196843 \\
\omega_3 = 0.080739 \\
\omega_4 = 0.134108 \\
\omega_5 = 0.143483 \\
\omega_6 = 0.396900
\end{matrix}
$$

[a] From Moffett *et al.* (2006). Reproduced with permission from Elsevier.

Sensitivity analysis

The two criteria with the largest weights were criterion 6, wildlife, and criterion 2, human population, and the results obtained by the single-dimensional sensitivity analysis of varying the weights assigned to them are shown in Figure 7.7. As was expected, given the similar composition of the alternatives, the rank orderings produced under such alternatives differed little from the final ranking.

A summary of the results obtained from the multi-dimensional rank order analysis based on the 10 000 rank orderings is provided in Table 7.3. Relatively little difference was found between these rankings and the final ranking. Overall, within these 10 000 rankings, alternative 7 was again identified as the preferred choice, based upon its mean and mode rankings. In the 10 000 rankings it was never ranked lower than second. The only other alternative to be ranked first was alternative 30, which was ranked second in the final ranking.

(a)

(b)

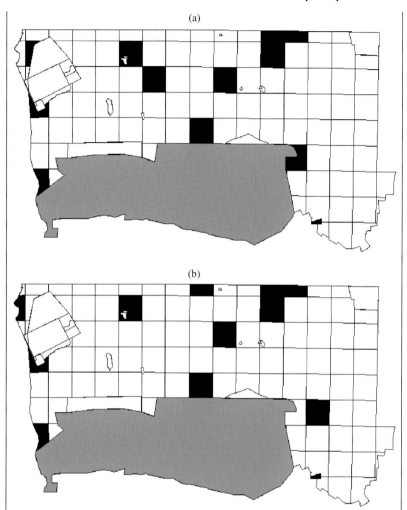

Figure 7.6. Conservation-area network for north-central Namibia. The maps show north-central Namibia starting with the Etosha National Park, the large planning unit in grey at the bottom of the map and going north to the border with Angola. The black cells represent planning units additional to the Etosha Park chosen for inclusion in a conservation-area network, while the white cells represent planning units not chosen. (a) Alternative 7 was ranked as the optimal network using the mAHP in conjunction with the comparisons shown in Table 7.3. (b) Alternative 93 was ranked fourth and is the alternative within the top five that exhibited the greatest amount of variation from alternative 7. Both alternatives contain 13 cells, 11 of which are shared. (From Moffett *et al.* (2006). Reproduced with permission from Elsevier.)

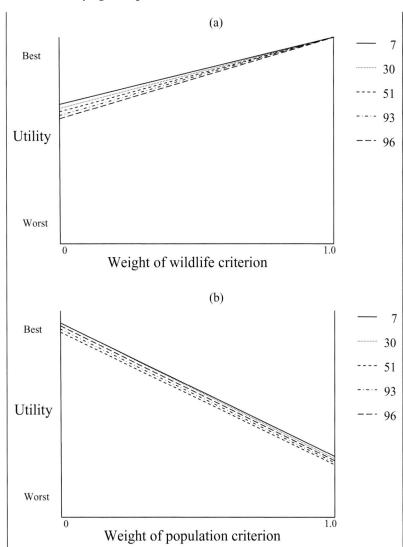

Figure 7.7. Sensitivity to changes in individual weights from the north-central Namibia case study. The effects of varying the weight associated with one criterion at a time, while holding the relative values of all other weights constant, was analyzed. Five alternative networks are illustrated, alternatives 7, 30, 51, 93 and 96 from Table 7.3. (a) demonstrates the effect on these alternatives of varying the weight associated with criterion 6, the number of wildlife and (b) provides a similar analysis for criterion 2, human population. (From Moffett *et al.* (2006). Reproduced with permission from Elsevier.)

Table 7.3 *Sensitivity to changes in multiple weights. Results from multi-dimensional rank order sensitivity analysis performed using the mAHP on 10 000 randomly assigned sets of criterion weights preserving the same rank ordering as the weights shown in Table 7.2. Column (1) lists one of the 49 non-dominated alternatives, columns (2) and (3) list the best and worst ranks of that alternative, columns (4) and (5) provide the mean and mode ranks of the alternative, while columns (6)–(10) provide the rank of the alternative at different percentiles[a]*

Alternative	Actual rank	Best rank	Worst rank	Mean rank	Mode rank	5%	25%	50%	75%	95%
1	38	34	41	37.51	38	35	37	38	38	39
2	8	5	22	8.66	8	7	8	8	9	12
3	15	8	20	14.46	15	9	14	15	16	16
5	18	14	20	17.58	18	17	17	18	18	18
6	9	7	13	8.87	9	8	9	9	9	10
7	1	1	2	1.335	1	1	1	1	2	2
8	14	7	18	12.79	14	8	12	14	14	15
9	46	44	49	46.84	47	46	47	47	47	48
13	17	15	23	17.62	17	17	17	17	18	19
15	16	14	22	15.69	16	14	15	16	16	17
18	45	40	48	45.65	46	44	45	46	46	46
19	10	8	21	10.78	10	10	10	10	11	13
20	24	20	27	22.57	24	20	21	23	24	25
21	47	46	49	48.31	48	47	48	48	49	49
23	29	10	33	26.82	29	17	25	28	29	32
27	27	20	41	29.44	33	24	27	29	32	36
30	2	1	4	1.666	2	1	1	2	2	2
36	49	12	49	41.17	49	29	37	42	46	49
38	33	28	35	31.65	31	30	31	31	32	34
39	11	5	32	12.37	11	10	11	11	14	16
42	44	24	46	42.98	45	39	42	44	45	45
46	35	27	37	33.86	35	32	33	34	35	36
48	7	3	18	7.299	7	5	7	7	7	11
50	26	20	33	26.97	26	26	26	27	28	30
51	3	3	13	3.015	3	3	3	3	3	3
53	32	28	46	34.18	35	29	32	35	35	40
54	20	2	31	21.31	20	19	20	21	23	24
56	40	35	47	41.68	41	40	41	42	43	44
59	6	6	14	6.175	6	6	6	6	6	8
60	39	34	43	40.47	40	39	40	41	41	42
63	28	24	30	27.21	28	26	27	27	28	28
64	23	20	30	22.34	20	20	21	22	24	25
66	42	36	45	42.56	43	41	42	43	43	44

(*cont.*)

Table 7.3 (*cont*)

Alternative	Actual rank	Best rank	Worst rank	Mean rank	Mode rank	5%	25%	50%	75%	95%
67	48	39	49	48.09	49	46	48	48	49	49
69	13	6	19	13.28	13	12	12	13	14	15
70	37	34	43	37.59	38	36	37	38	38	39
74	43	41	48	44.04	44	43	43	44	45	45
79	31	12	39	29.93	30	26	29	30	32	33
81	22	20	26	22.06	22	21	21	22	23	23
82	25	22	34	26.34	25	25	25	26	27	30
83	41	14	43	38.99	39	33	39	39	41	42
84	21	3	32	22.62	21	20	21	23	24	25
89	19	16	22	19.09	19	18	19	19	19	21
91	34	11	37	31.61	34	27	30	32	34	35
93	4	3	11	4.003	4	4	4	4	4	4
94	12	7	26	11.18	12	7	10	12	13	13
96	5	4	12	5.058	5	5	5	5	5	6
98	30	23	40	32.76	34	30	31	33	34	38
99	36	35	40	36.52	36	36	36	36	37	38

[a] From Moffett *et al.* (2006). Reproduced with permission from Elsevier.

7.6 Summary

Conservation planning must incorporate into the selection of conservation area networks criteria other than the targeted representation of biodiversity surrogates. These criteria include spatial design criteria and socio-political criteria. The importance of the latter lies in the fact that conservation planning always occurs in a socio-political context, in which designating areas for biodiversity conservation must compete with other claims on natural resources.

A variety of techniques has been developed for multi-criteria analysis (MCA) and many of these have been used for conservation planning. There are two protocols for the use of MCA techniques: (1) they can be used when each area is individually selected for inclusion in a potential conservation area network or (2) a variety of networks can be selected to satisfy conservation goals and MCA techniques can be used to discriminate among them. Both protocols can also be used simultaneously during a planning exercise.

Many MCA methods require precise information on the value of each alternative option according to each criterion and a precise ranking of the criteria, both of which are difficult to obtain in practice. MCA should begin with the computation of the non-dominated set (NDS) of alternatives because selecting these requires the least information. If the NDS is too large, trade-off analysis and a modified analytic hierarchy process are two standard techniques available to refine that set.

8 · *Systematic conservation plans*

In this chapter we describe five examples of systematic conservation plans that have been developed in different parts of the world. Each one took place in a different ecological setting and so different biodiversity surrogate sets were used. Just as importantly, each one differed in the resources available for data collection, treatment and planning as well as in the cultural, social and economic contexts. Therefore, while there are commonalities in the systematic approach, there are differences in the problems that needed to be solved in order to come up with a plan. In this respect, the process of coming up with a plan is as useful a product as the plan itself. Using conservation biology as an example, and noting similarities with medicine, Sarkar (2004) wonders if in future the most valuable science will consist of building highly contextual procedures rather than producing or seeking general theories. We believe that the process of deriving each of the conservation plans described below, including the data collection and treatment stages described earlier in Chapters 2, 3 and 4, is at least as valuable as the plan itself, especially given that conservation planning must be an ongoing iterative process if it is to be, as it must be, incorporated along with social, economic and political considerations in regional natural resource management processes.

8.1 Complementarity by inspection in the Nullarbor region, Australia (Plates 15 and 16)

Following the survey in the Nullarbor region and treatments of the data collected in that survey (Chapters 3 and 4; Examples 3.3 and 4.2), there was a matrix of 80 quadrats by 14 species assemblages, in which each value referred to the proportion of the assemblages' species that was present at each quadrat. The proportional richness of 13 of the 14 assemblages was interpolated across the region and plotted. The remaining assemblage, number 7, comprised only two species and had a disjunct distribution as a result of which no consistent pattern could be discerned. Figure 8.1

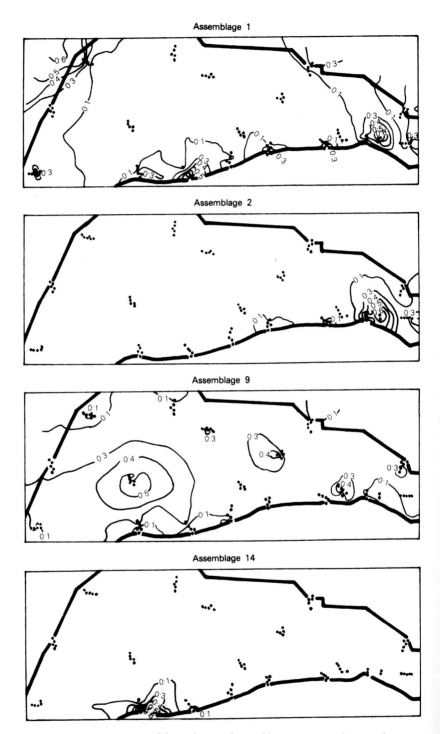

Figure 8.1. Contour maps of the richness of assemblages 1, 2, 9 and 14 on the Nullarbor Plain as examples. Contours represent the proportion of the assemblage's total richness. (From McKenzie *et al.* (1989). Reproduced with permission from Elsevier.)

shows these plots for assemblages 1, 2, 9 and 14 as examples. Table 8.1 is a part of the two-way table derived from the classification explained in Chapter 4. It describes the compositional variation within three of the assemblages. Using the interpolated richness plots in conjunction with this table, it was possible to implement a form of complementarity by inspection to identify potential new conservation areas, as described below.

The examination of the complete two-way table revealed that assemblages 1 and 9 are widespread, have more than one node of richness and exhibit gradients in species composition. For example, the 50% of assemblage 1 species in the south-east node of richness (Figure 8.1) is not the same set of species as the 50% in the south-central node, although there is some overlap. To a lesser extent this was also true of assemblages 8, 10 and 12. Assemblages 2, 3, 4, 5, 6, 11, 13 and 14 are localized with a single node of richness that encompasses their entire species composition. By overlaying the maps of compositional richness and with the aid of the full two-way table, it was possible to identify areas likely to contain a range of assemblages as well as areas with few assemblages that may not occur elsewhere (Figure 8.2). The limit of the resolution of the data was the size of the sampled quadrats ($4\,km^2$). In some cases, only three or four such quadrat-sized areas were needed to encompass the compositional range of an assemblage. For example, the area around one quadrat near campsite BA encompassed all species of assemblage 13, and the area around three quadrats near campsite MA encompassed assemblages 5 and 6. In other cases, areas were required in several locations. For example, areas in the vicinity of campsites HA, HU and CA were required to encompass the compositional variation of assemblage 9.

Six protected areas already exist within the Nullarbor region and cover 14% of its area. The compositional variation of only 6 of the 14 assemblages is encompassed by that existing network. Figure 8.3 shows those existing reserves as well as new areas proposed as a result of this survey (Robinson *et al.* 1987). These new areas would encompass an additional six assemblages and could be located on vacant land belonging to the state. The compositional variations of assemblages 2 and 13 occur only on pastoral properties (ranches) and so it would be necessary to purchase these, or come to a management arrangement with the owners, if all 14 assemblages were to come under some form of conservation management.

McKenzie *et al.* (1989) concluded that the procedure they followed was less subjective than other conservation area selection methods because:

Table 8.1 *Part of the two-way table derived from the classification of assemblages explained in Chapter 4. Species composition within assemblages is displayed in the rows*[a]

Site Groups

Species Assemblage	1	2	3	4	5	6	7	8	9	10	11	12	13	14	15
	BBBBCKMKK	IIIKKKKK	FFFF	FMMMMMJJJJPPPP	CCCCC	KY	KKKKM	HHHHH	HHHHII	H	BCMMM	CCC	KM	KMMMM	YYYY
	AAAAOUAUU	FFFFOOOOO	OOOO	OUUUUUUUULLLL	AAAAA	DA	DDDUE	AAAAA	UUUUFF	U	AOAAA	OOO	UA	DEEEE	AAAA
	124512143	34513452	1243	5123451234512534	12345	15	34515	12453	123512	4	32234	345	55	21234	1243
Acanapic	** *										***	***	*****	***	** *
Phylalbi	*	*		**							***	***	****	***	** *
Smichbrev	*	*	*	*							*****		*	***	***
Exocaphy		****		*							****	*		****	****
Micrleuc	*	*		**							****			*	**
Pardstri	**	*		*							******	*	*	***	***
Eucaoleo	*	*		*							****		*	**	**
Oleamuel	*	*		**							****			***	***
Scaespin				*				*					*	****	****
Dodosten				*							*		**	*	*
Eucasoci				*									*	****	**
Aegocris				*							***		*		*
Amphmino	*	*		****							*			*	**
Cherleuc			*	*					*		***	*			
Artapers	****			****							***				
Geijparv	****	*		**						*	****	**			
Crypplag		*		*										*	

Acanired

Lyciaust

Tympline

Moreadel

Scleobli

Anthnova

Undemili

Cinccrur

Smincras

Falccenc

Mairsedi

Salskali

Hiruneox

Amphnull

Atrinumm

Ardeaust

Ctenuber

Arabtris

Sclepate

Atriacut

Ephtalbi

Nitrbill

Cincalis

(cont.)

Table 8.1 (*cont.*)

Atrieard
Planbell
Circassi
Ephtauri
Peltaust
Mairturb
Eremlong
Tillocci
Atricryp
Eremmacu

Lichcrat
Egermult
Adenforr
Beaumicr
Borocras
Conodrum
Gahnlani
Lepivisc
Loxoflex
Lysicili
Pomamyrt
Malupulc

14

Caliprei				***
Dodovisc				***
Eucaincr				***
Acaccoch				**
Acacniti				**
Eucadive				*
Eucafoec				*
Pultobco				*
Amphmacu			*	***
Leridist			*	***
Hakeniti			*	***

Assemblages 1 and 9 exhibit compositional gradients, whereas assemblage 14 has all species confined to a small area. The letters and numbers under the site groups refer to the campsites and associated quadrats shown in Fig. 3.10 in Chapter 3. Reading down the first column, BA1, BA refers to campsite BA and 1 refers to Quadrat 1 of that campsite. Abbreviations of species names are used at the beginning of each row. On the first row, Acanapic is an abbreviation of *Acanthiza apicalis*, inland thornbill. On the first row of assemblage 9, Acanired refers to *Acanthiza iredalei*, samphire thornbill, and for assemblage 14, Lichcrat refers to *Lichenostomus cratitus*, purple-gaped honeyeater. The last species listed in each assemblage are Crypplag, *Cryptoblepharus plagiocephaulus*, a small skink, Eremmacu, *Eremophila maculata*, native fuschsia and Hakeniti, *Hakea nitida*, shining hakea.

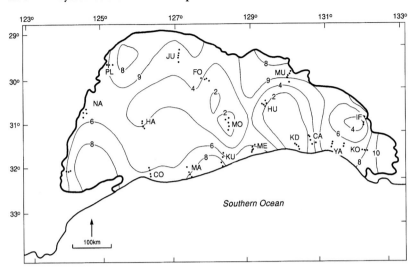

Figure 8.2. Centres of assemblage richness on the Nullarbor Plain derived by overlaying individual assemblage maps. For example, eight assemblages occur near campsite MA but only three near HU. (From McKenzie *et al.* (1989). Reproduced with permission from Elsevier.)

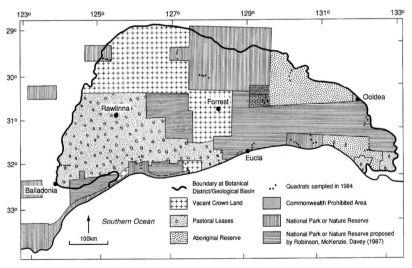

Figure 8.3. Land tenure in the Nullabor region, showing existing and proposed conservation areas. (From McKenzie *et al.* (1989). Reproduced with permission from Elsevier.)

(1) It allowed more precise identification of gaps in the conservation-area network for a wide variety of ecologically different organisms through the individual assemblage models;
(2) It identified areas that would achieve representation of the compositional variation within each assemblage;
(3) It allowed them to select from among a variety of options, the new areas that would improve coverage of assemblage compositional variation.

8.2 Complementarity using species records in Québec

In the early 1990s, Canada's federal and provincial environment, parks and wildlife ministers all signed a statement of commitment to an "Endangered Spaces Campaign" which proposed that at least 12 % of each natural region in the country be put under some form of protection. It was generally assumed that the target of 12 % would be achieved individually in each province (Hummell 1995). In 1992, the Québec provincial government officially announced its intention to complete a representative network of protected areas by 2000. In 1989, about 6090 km^2 or 0.4 % of terrestrial Québec was under some form of protection. By 1998 this had increased to 63 960 km^2 or 4.2 % of the land area.

Sarakinos *et al.* (2001) developed a nominal conservation area network for Québec which was briefly discussed in Chapter 5 (Example 5.5). Their goal was to assess the representativeness of the existing protected areas and the new ones that had been proposed, as well as use complementarity to identify areas of high priority. They also compared these high priority areas with the only other explicit plan that had been produced, by the World Wildlife Fund, Bureau de Québec – Union québécoise pour la conservation de la nature (WWF-UQCN 1998).

The province (approximately 1 522 842 km^2) was divided into 21 403 cells at a 0.2° × 0.2° longitude × latitude resolution which comprised the planning units for this analysis. The average cell size was about 74 km^2. As (true) surrogates for biodiversity they used 400 faunal and floral species at risk (346 plant species and 54 animal species), 22 native small mammals, 6 game mammals, and 92 fish species. The category "at risk" was defined by the former Québec Ministry of the Environment and Wildlife. Data on the distribution of species at risk were obtained from the Québec Natural Heritage Data Centre of this ministry. The database consisted of point locations for 346 of 374 plant species at risk and for 48 of 78 animal species at risk. Locations for more threatened and endangered species were withheld because of the sensitive nature of this information. However, for eight of these other animal species, other agencies provided public distribution information. The absence of some of the data on species at risk was the major lacuna of this analysis. However, it was presumed that individual conservation plans were being devised for these species and, therefore, their habitat would be protected whether or not it was included in the priority areas that were identified. Small and game mammal and fish data came from other governmental agencies. The ready availability of so much georeferenced data on species at risk distinguishes this analysis from the others discussed in this chapter.

The data were a mixture of presence–absence and presence-only records. No data treatment was attempted. The rationale was that, for species at risk, false presences, as would at least occasionally be generated by modeling treatments, constituted an unacceptable risk. All data were treated as presence–absence. Once again, the rationale was precautionary: if areas selected also included presences of unrecorded species at risk, these would only be even better represented in the conservation area network. This virtue was presumed to outweigh the possible loss in economy because the absence of records of species in cells was being interpreted as their absence on the ground. A separate analysis was also carried out for southern Québec because bird distribution data were also available only for that region – it will not be further discussed here.

As noted in Chapter 5 (Section 5.4), Figure 5.4 shows that, except in southern Québec, species at risk are represented poorly in the existing protected areas. Figure 5.5 shows the areas selected using the algorithm discussed there with a target of 50 representations for each of the species at risk, and 100 representations for the other species. Only six species had higher representation levels and including all of these did not produce significantly different results. As also noted in Chapter 5 (Section 5.4), Sarakinos *et al.* found that the representation of species at risk within the existing protected areas was so poor that results of selection initiated with the existing protected areas and selection *ab initio* did not lead to significant differences. They obtained two other interesting results:

(1) In Figure 8.4 the arrow shows a band of selected cells in the northwest. This band only begins to be selected when targets of representation are set at all records for species at risk and 200 representations for all other species. They become continuous at a target of 250 representations for these other species, which is the case depicted in Figure 8.4. This band runs through boreal forest, which is the most northerly

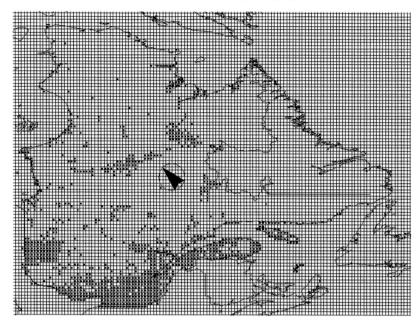

Figure 8.4. The cells selected in Québec for representation of all species at risk and at least 250 representations of other species. The arrow shows the continuous band of boreal forest. For more details, see the text. (From Sarakinos *et al.* (2001). Reproduced with permission from Kluwer Academic Publishers (Springer Science).)

and abundant of Québec's three forest zones and straddles the Canadian Shield and upper lowlands region of the province. Because of the milder climate, the diversity of organisms is also higher, with approximately 850 plant species and 281 vertebrates. The boreal forest covers 27 % of Québec and only 0.1 % is under any legal protection though 87 % of it is owned by the provincial government (Thibault 1995). The main threat to it is due to logging; most logging in Québec occurs in this ecozone and about 80 % of logging in Québec still uses clear-cutting methods. Logging contributes only 1 % to Québec's GDP even though about 0.5 % of the forests are logged annually producing about 80 000 jobs. In the 1990s logging of the boreal forests emerged as an issue of public concern. This analysis showed that, while these concerns were partly justified, since these areas are not selected when species at risk are used as surrogates, logging could be phased out gradually over a period of years, allowing alternative employment to be found for the affected people, without endangering

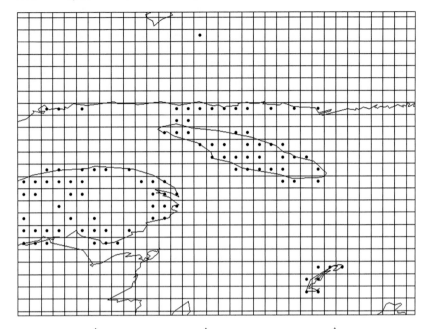

Figure 8.5. The Île d'Anticosti and the Îles-de-la-Madeleine. The Île d'Anticosti is in the middle of the figure and the Îles-de-la-Madeleine is at the bottom right of the figure. The selected cells are shown in black. The targets were 50 representations of species at risk. (From Sarakinos *et al.* (2001). Reproduced with permission from Kluwer Academic Publishers (Springer Science).)

species. However, this conclusion may not hold because of the limitations of the data used: some species at risk were not included and presence-only data were interpreted as presence–absence records;

(2) The Île d'Anticosti and the Îles-de-la-Madeleine emerged as areas of particular conservation interest (see Figure 8.5). In these islands, which have very sparse human use, conservation may be easily achieved by the creation of conventional protected areas with limited human access. In the Île d'Anticosti, a significant part of the coast and two coast-to-coast strips were selected. The Parc de la Rivière-Vauréal, which had been proposed in the 1990s and was under public consultation, only included the northern cells. The analysis – and WWF-UQCN had reached similar conclusions using less systematic methods – showed that the proposed park should be bigger. In the Îles-de-la-Madeleine the analysis selected the entire island even when the target was only one representation of each species at risk. Thus this island was in serious need of protection. The proposal from the WWF-UQCN had not included the entire area.

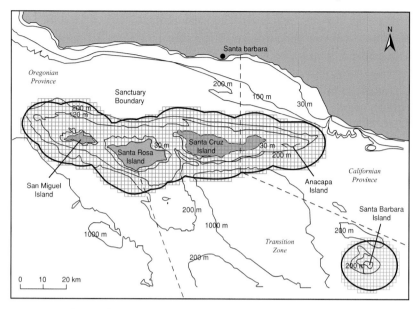

Figure 8.6. The California Channel Islands Planning Region. Each planning unit was about 1′ × 1′ longitude × latitude (roughly 1 × 1 nautical miles). Each unit was assigned to a depth interval: Photic zone (0–30 m), shallow continental shelf (30–100 m), deep continental shelf (100–200 m) and continental slope (>200 m). (From Airamé *et al.* (2003). Reproduced with permission from the Ecological Society of America.)

8.3   A marine conservation plan for the California Channel Islands, United States (Plates 13 and 14)

The examples we have discussed so far are all terrestrial. However, the principles of systematic conservation planning discussed in this book can be applied without change to many marine conservation contexts. In California, in 1999, a process to develop a marine conservation-area network was initiated and a body called the Marine Reserves Working Group (MRWG) was formed. This group included officers of federal and state agencies, commercial and recreational fishermen, environmentalists and other members of the Santa Barbara community. The work of the group was assisted by a Science Advisory Panel (SAP) that evaluated ecological and physical data and conducted the analyses described below and a socio-economic panel that evaluated economic data. The study was restricted to the Channel Island National Marine Sanctuary, which prohibits exploration for gas and oil, but allows both commercial and recreational fishing (see Figure 8.6). Different areas within the sanctuary were under different regulatory regimes. The goal was to identify a set of marine reserves that would: (1) protect ecosystem biodiversity, (2) help maintain sustainable fisheries, (3) contribute to long-term economic viability, (4) maintain the natural and cultural heritage of the sanctuary and (5) foster stewardship of the marine environment by creating educational opportunities. Full details of this process and the analyses summarized below can be found in Airamé et al. (2003).

The planning region was divided into about 1500 planning units, each of size $1' \times 1'$ longitude × latitude (Figure 8.6). Boundaries of planning units were irregular where they intersected with the shore or sanctuary boundary. This scale was chosen because it was the one at which socio-economic information on commercial and recreational use had been collected. The "cost" of each planning unit was its size, measured by its boundary. (Socio-economic costs were not used in the area selection process, but they were taken into account post-analysis by the MRWG in settling on a final network.) The region was divided into three biogeographical zones (see Figure 8.7). The biodiversity surrogates used were 39 habitat types defined by physical variables as well as plant communities, and seabird and pinniped habitat (see Table 8.2). For each surrogate, the target was set to 30–50 % of its extent with the aim of protecting 30–50 % of the entire region.

The analysis did not take existing reserves into account to determine priority areas. If existing reserves did not overlap with the selected priority areas, they were regarded as tradable. Area selection was done using SITES (Andelman et al. 1999; Possingham et al. 2000), which uses the simulated annealing algorithm (Kirkpatrick et al. 1983) to choose areas. The goal was to select potential sets of 1–4 reserves separately in each of the three biogeographical zones to represent the surrogates adequately while minimizing the total selected area (or cost). The algorithm makes a random selection of planning units followed by an assessment of the performance of this selection in achieving the stated goals, then randomly selects planning units, successively, and evaluates the change in goal achievement with the addition or deletion of each planning unit. At each step, the new solution is compared with the previous one and the one closer to achieving the goal is accepted as input to the next iteration. Thus, each run produces a different solution. It was run 300–800 times depending on the size of the biogeographical zone. Spatial clustering of planning units was encouraged by giving a preference to adjacent cells during the selection process and by attempting to minimize the total boundary length, a distinctive feature of SITES and related software packages

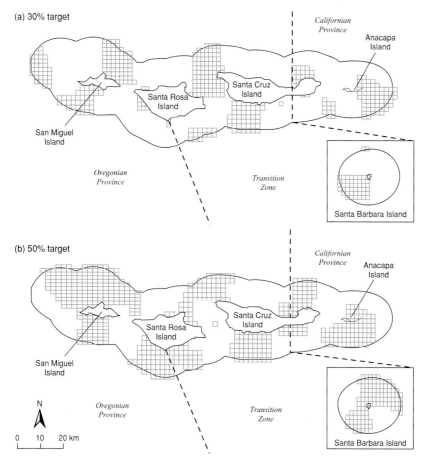

Figure 8.7. Areas selected for a conservation plan in the Channel Islands. (From Airamé *et al.* (2003). Reproduced with permission from Ecological Society of America.)

(e.g., MARXAN [Ball & Possingham 2000]). Figure 8.7 shows the results obtained when 30% and 50% of the total area is targeted for protection. The irreplaceability value of a selected planning unit was measured by the frequency with which it was selected in the runs of the algorithm.

At the end of the process, the SAP presented the MRWG with the ten best spatially explicit conservation area network scenarios (Airamé *et al.* 2003). Having more than one such alternative allows flexibility during implementation. Thus, the MRWG was able to evaluate alternative networks in an interactive GIS that included summed irreplaceability

Table 8.2 Surrogates for the California Channel Islands[a]

Ecological criteria	Units	Oregonian province	Transition zone	Californian province
Coastline characteristics				
Sandy beach	km of coastline	39.7	22.1	7.5
Rocky coast (low exposure)	km of coastline	45.1	18.6	20
Rocky coast (high exposure)	km of coastline	43.8	21.8	2.2
Substrate type and depth				
Soft sediment (0–30 m)	km^2	133.4	101.5	56.3
Hard sediment (0–30 m)	km^2	117.6	24.7	22.6
Soft sediment (30–100 m)	km^2	725.8	218.1	192.8
Hard sediment (30–100 m)	km^2	80.3	34.6	13.4
Soft sediment (100–200 m)	km^2	538.5	215.7	93.3
Hard sediment (100–200 m)	km^2	–	25	3.8
Soft sediment (>200 m)	km^2	777.6	606.7	551.2
Hard sediment (>200 m)	km^2	–	50.1	7.9
Additional features				
Emergent rocks (nearshore)	no. <2 km from shore	216	208	95
Emergent rocks (offshore)	no. >2 km from shore	12	5	1
Submerged rocky features (pinnacles, ridges, seamounts)	km^2	20.2	91.6	13.7
Submarine canyons	km^2	3.4	115.6	17.1
Dominant plant communities				
Giant kelp	km^2	55.2	20.2	6.2
Surfgrass	km^2	46	23	11
Eelgrass	km^2	1	0.3	0.7

Breeding seabirds				
California brown pelican	km of coast[b]	0.5	10.1	9.4
Pelagic cormorant	km of coast[b]	115.7	39.7	37
Double-crested cormorant	km of coast[b]	0.5	0	8.5
Brandt's cormorant	km of coast[b]	66.1	11	23.2
Common murre	km of coast[b]	0.5	10.1	0
Pigeon guillemot	km of coast[b]	76.6	29.6	39
Xantu's murrelet	km of coast[b]	8.5	13.8	15.4
Tufted puffin	km of coast[b]	0.5	10.1	0
Rhinocerous auklet	km of coast[b]	8.5	0	0
Cassin's auklet	km of coast[b]	21.1	14.2	7.4
Leach's storm-petrel	km of coast[b]	12.2	10.1	4.2
Ashy storm-petrel	km of coast[b]	1.1	0	0
Black storm-petrel	km of coast[b]	0.5	10.1	0
Black oystercatcher	km of coast[b]	121.4	49.6	40.6
Snow plover	km of coast[b]	28.5	9.4	0
Western gull	km of coast[b]	85.0	25.9	32.8
Pinnipeds				
Stellar (northern) sea lion	km of coast[c]	6.4	0	0
California sea lion	km of coast[c]	9	10.1	0
Northern fur seal	km of coast[c]	6.9	0	0
Northern elephant seal	km of coast[c]	28	6.4	0
Harbor seal	km of coast[c]	76	40.2	20

[a] From Airamé et al. (2003). Reproduced with permission from the Ecological Society of America

[b] Kilometers of coast suitable for nesting

[c] Kilometers of coast used by pinnipeds for hauling out

values for each planning unit as well as socio-economic information on major commercial and recreational activities and in this way perform what amounted to a terminal-stage multi-criteria analysis (Chapter 7, Section 7.1).

8.4 A conservation plan for the Cape Floristic region of South Africa (Plates 17 and 18)

The Cape Floristic region (CFR), an area of $87\,892\,km^2$ at the south-western end of South Africa, is a widely recognized centre of endemism for both plants (Goldblatt & Manning 2002) and animals (e.g., Branch 1988; Picker & Samways, 1996). It is also a global biodiversity hotspot (Myers *et al.* 2000). Between 1998 and 2000, a systematic conservation plan and implementation strategy for the entire CFR was developed. It was known as the CAPE (Cape Action Plan for the Environment) project and consisted of four stages: (1) situation assessment (including a conservation planning component), (2) strategy development, (3) action-plan formulation and (4) fundraising. Cowling and Pressey (2003) provide an overview of the project and full details have been published in a series of papers comprising a special issue, Volume 112, Nos. 1–2, 2003, of the journal *Biological Conservation*. Here we summarize the conservation planning component drawing on material from Cowling and Pressey (2003), Cowling *et al.* (2003), Kerley *et al.* (2003), Lombard *et al.* (2003), Pressey *et al.* (2003), Rouget *et al.* (2003a) and Rouget *et al.* (2003b).

The CFR has a long history of reserve establishment going back to the late 1800s. By the late 1990s, although more than 20 % of its area was under some form of conservation management, like most other places in the world (Chapter 5; Pressey 1994) it was not representative of the CFR's biodiversity. At this stage, the CFR was faced with escalating threats from agricultural transformation, woody weed infestation, urbanization and, most recently, harvesting of products of the dominant vegetation type, the fynbos, such as bush teas and cut flowers, as well as a diminishing institutional capacity for management (Cowling & Pressey 2003). In attempting to address these issues, the CAPE project adopted a systematic conservation planning approach, which followed a planning protocol based on those in Margules and Pressey (2000) and Pressey and Cowling (2001). These protocols have been further developed since, and are presented in this book in Table 1.1 of Chapter 1. In keeping with their protocol, Cowling and colleagues first identified key stakeholders and involved them in the process from the beginning. They also analyzed the policy, legal, institutional and socio-economic environments in which they were operating in order to identify opportunities for and constraints on implementing conservation plans.

8.4.1 Biodiversity surrogates

The next step was to evaluate available data on biodiversity features and threats in order to choose biodiversity surrogates (Lombard *et al.* 2003; Pressey *et al.* 2003). Table 6.3 (Chapter 6; Section 6.3) shows the surrogates that were used. They are a combination of habitat types, species and spatial patterns of processes. Broad habitat units represent intuitive land classifications based on the similarity between areas of climate, geology, topography and vegetation type. Data on Proteaceae were available at a consistent level of detail (low level of spatial bias) across the whole CFR, so Proteaceae species were used. Field records of vertebrate species were also available for the whole CFR, although more spatial bias was exhibited by the 8840 records of 345 species (121 fish, 48 amphibians and 176 reptiles). Kerley *et al.* (2003) estimated the number of individuals needed for functional social or breeding groups, or for effective management, of 41 large- and medium-sized mammals many of which are charismatic and iconic African species. These estimates were based on extensive scientific knowledge and expert opinion. The inclusion of surrogates for ecological processes is an attempt to ensure that conservation areas encompass functions that are necessary to sustain biodiversity into the future. They include ecological gradients in altitude and soil type as well as water-catchment processes. This was the first attempt explicitly to incorporate processes into systematic conservation planning. For edaphic interfaces and upland–lowland interfaces they mapped a 1 km buffer, 500 m each side of the boundaries and excluded those parts of these interfaces that were transformed by urban development, agriculture and/or invasion of exotic species. For sand movement corridors they mapped each corridor, excluded areas irretrievably lost to urban development, agriculture or invasive species and retained the other portions of these corridors. For interbasin riverine corridors they mapped 500 m wide buffers (250 m on either side) and then excluded those portions that were transformed by urban development, agriculture or invasive species. Margules and Stein (1989) discussed the benefits of locating reserves along the steepest ecological gradients, in their case altitude (itself a surrogate for temperature) and rock type. They also identified management issues which included, in their study area of south-eastern New South Wales, Australia, the fact that major infrastructure, such as roads and powerlines, ran perpendicular to this gradient and therefore cut it in a number of places. However, unlike the CAPE project, they did not attempt to map the spatial extent of these gradients or include them explicitly in a conservation plan. If variables representing key processes can be identified and mapped they can and should be incorporated into conservation-area planning.

8.4.2 Planning units and targets

The next steps were to delineate planning units and create targets for the surrogates. Planning units are the candidate conservation areas. Cowling *et al.* (2003) used ~3900 ha grid cells. Existing protected areas were included as commitments (see Section 8.5 below) so that their contributions to biodiversity goals could be counted, but they were not candidates for selection (or de-selection). There are two kinds of protected areas in the CFR, statutory reserves and non-statutory reserves. Statutory reserves became planning units but non-statutory reserves were subdivided by the grid cells where the boundaries crossed cells, enabling non-statutory reserves to become candidates for selection. Only those parts of non-statutory reserves that contributed most to targets were selected. They also incorporated the boundaries of fixed process surrogates (edaphic interfaces, riverine corridors, sand movement corridors and upland–lowland interfaces) into the planning unit layer, intersecting them with the grid cells. They did not include flexible process components (upland–lowland gradients and macroclimatic gradients) since there were spatial options for incorporating them into a plan. Altogether, there were 7039 planning units: 3014 grid cells, 2993 process planning units and 1032 protected areas. Conservation targets were created for each surrogate, as described in Chapter 6 and, in more detail, in Pressey *et al.* (2003).

8.4.3 Building the conservation plan

The objective was to meet all targets for all surrogates in a set of planning units, or conservation areas. The plan was developed in seven stages, described in detail by Cowling *et al.* (2003). The first four incorporated planning units for which there were no options if the targets were to be met. The second three incorporated planning units for which there were options. At each stage, the contributions of the selected planning units to meeting all targets were measured and recorded, so that in subsequent stages those contributions could be taken into account. The

C-Plan software package (Pressey 1999; Ferrier *et al.* 2000) was used to support the planning process and assess the contributions of selected planning units to the conservation targets. In the first stage, statutory reserves were committed and the contribution they made to all targets measured. In the second stage, areas comprising untransformed parts of edaphic and upland–lowland interfaces, sand movement corridors and interbasin riverine corridors not already covered in statutory reserves were committed. In the third stage, all planning units that had to be in the plan if the species-based targets were to be achieved, that is, those areas that were irreplaceable if the targets for Proteaceae, vertebrates and habitat units were to be met and had not already been selected in the first two stages, were committed. In the fourth stage, areas identified by Kerley *et al.* (2003) as necessary for meeting targets for large- and medium-sized mammals, and had not been covered in previous stages, were committed. In the fifth stage, areas needed to complete representation of the macro-climatic gradients, that is, those parts of the gradients not already selected incidentally in previous stages, were selected. In the sixth stage, areas needed to complete representation of the upland–lowland gradients, that is, those parts of the gradients not already selected incidentally in previous stages, were selected. In the seventh stage, planning units needed to achieve all outstanding targets for habitat units, Proteaceae and vertebrates were selected using a measure of complementarity (called irreplaceability) and lowest level of threat to break ties. Figure 8.8 is a map of the CFR showing how the plan was built up progressively, through the seven stages.

The final conservation plan for the CFR covered 52.3 % (49 958 km^2) of the remaining natural habitat in the region. Statutory reserves comprised 10.1 % of the CFR, so the additional area demanded by this plan was 42.2 %, approximately 40 000 km^2. The spatial configuration of the plan was more strongly influenced by some surrogates than others. The existing statutory reserves over-achieved targets for many montane habitat units, achieved the targets for most vertebrates and Proteaceae species and, because they were the starting point, constrained the economy with which large- and medium-sized mammal populations could be represented. The result was that large areas with low overall complementarity were needed to achieve these targets, leading to very high over-achievement of many species and habitat unit targets. All remaining untransformed habitat was needed to achieve targets for 12 highly threatened habitat units. The spatial configuration of this plan is only one of many alternatives for achieving the set targets. While many planning

218 · **Systematic conservation plans**

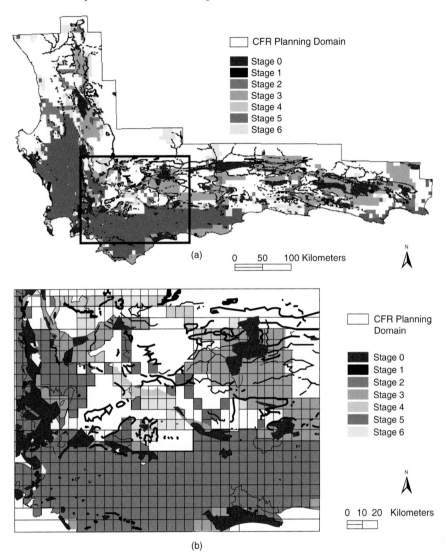

(a)

(b)

Figure 8.8. The Cape Floristic Region. Showing all (a) and an enlarged inset area (b) of a proposed progressive system of conservation areas that would achieve targets for all biodiversity features in seven stages, building on the existing reserves (stage 0). (From Possingham *et al.* (2006), redrawn from Cowling *et al.* (2003). Reproduced with permission from Elsevier.)

units are irreplaceable if targets are to be achieved, there are options for mammal populations and the upland–lowland and macroclimatic gradients. Thus, there is some flexibility to take account of other social goals such as the sustainability of livelihoods. Cowling and colleagues did not use multi-criteria analysis to incorporate the many variables they were faced with, particularly the social and monetary cost variables. However, the plan does provide the spatially explicit guidelines needed to achieve the goals of the Cape Action Plan for the Environment (CAPE) and its implementation would go a very long way to protecting the biodiversity of the Cape Floristic region.

8.5 A conservation plan for Papua New Guinea (Plates 3 and 4)

Following the collection and treatments of data described in Chapters 2 and 3, the biodiversity data for conservation planning in PNG consisted of 4470 map units, the candidate conservation areas, and 1183 biodiversity surrogates made up of 608 environmental domains, 564 vegetation types and 11 rare and threatened species. The goal was to represent each of these surrogates in a network of conservation areas. The approach adopted incorporated opportunity costs in order to try to balance conflicting demands on natural resources (Faith *et al.* 1996). Specifically, the objective in PNG was to find a set of conservation areas that together reached the conservation goal but minimized forgone opportunities for timber production. In searching for this set of areas, it was also planned to incorporate existing protected areas, avoid areas of high land-use intensity and human population density and prefer, where possible, that they coincide with areas chosen previously by experts as conservation priority areas. The following description is summarized from Nix *et al.* (2000) and Faith *et al.* (2001a).

Each candidate conservation area or planning unit, which were resource mapping units (RMUs, Chapters 3 & 4), contained one or more biodiversity surrogate. These surrogates had some quantitative value attached to them, for example, hectares of a vegetation type. Each was assigned a target, in this case a single representation of each surrogate, which is why small occurrences were masked out (see below). When implementing cost trade-offs, the complementarity value of each area was compared to its cost. An area was justified for selection if its complementarity value, its marginal contribution to overall biodiversity representation, exceeded its weighted cost. Areas were iteratively added and deleted from a select list so as to approach the target level of representation. Complementarity and cost are measured in different units so that to begin the process it was necessary to find a weight for cost that would allow, when the comparison was made, at least some areas with high complementarity to be admitted to a select list. This list was saved and the weight lowered so that more areas were admitted to the list. This process was repeated iteratively until the target had been achieved. In this way, the biodiversity goal was met and opportunity costs were minimized. The Target software package (P. A. Walker & D. P. Faith, personal communication) was used to find the set of planning units which achieved the conservation goal and minimized opportunity costs. One cost (timber volume) was therefore incorporated into the PNG planning process. There were data available for other costs, for example, agricultural potential, which could be incorporated if the study was being done now, using multi-criteria methods such as those described in Chapter 7. Five kinds of data were used to guide conservation-area selection in PNG. The first were the biodiversity surrogates, the environmental domains, vegetation types and rare and threatened species, already described in Chapters 3 and 4 and summarized above. The others are described below.

8.5.1 Opportunity costs

Index of timber volume
The forest inventory mapping system (FIM) described in Chapter 3 contains data on estimated timber volumes for each map unit. These were made available as an index ranging from zero for low volume to five for high volume. It was possible to aggregate these values for each resource mapping unit (RMU) and thus derive an estimated index of timber volume for each of these RMUs, the areas that were to be assessed as candidate conservation areas. This was not an estimate of timber volume per unit area and it meant that larger RMUs may have had high timber volume estimates even when volume per unit area was low. Figure 8.9 is a map of RMUs with the two highest classes of estimated timber volume.

Agricultural potential
The PNG Department of Agriculture and Livestock provided a simple model of agricultural potential. RMUs with slopes from less than 2° up to 20° that were well to imperfectly drained were designated to have agricultural potential and the rest were not. While agricultural potential was not used to derive the conservation priority areas, it was used to diagnose the results.

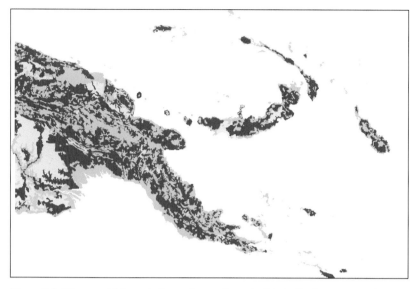

Figure 8.9. The two highest timber volume classes in PNG. Darkest grey is the second highest timber volume class, next darkest is the highest timber volume class and lightest is all other areas. The map units are resource mapping units (RMUs) from the PNGRIS database (see Chapter 3, Example 3.1), which were mapped as distinct patterns from aerial photographs. This is why they are so convoluted. (From Faith *et al.* (2001a). Reproduced with permission from Surrey Beatty & Sons Pty Ltd.)

8.5.2 Commitments

These were RMUs committed to the priority set regardless of their complementarity value. Existing protected areas were the only commitments used in this study. While there was considerable debate about the wisdom of doing this, since many are not performing the protection function for which they were established, it was deemed necessary if the results were to have credibility with government agencies. Figure 8.10 is a map of existing protected areas in 1998. RMUs with rare and threatened species would be other candidates for commitment instead of these species being used, as they were here, as surrogates for the calculation of complementarity.

8.5.3 Masks

Masks were used to exclude some RMUs from consideration regardless of their complementarity value. This was done in an attempt to address the vulnerability issue by ensuring that small and degraded areas could

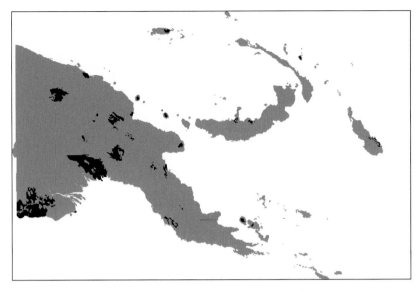

Figure 8.10. Existing protected areas in PNG in 1998. The boundaries were adjusted to fit RMUs, which is why they are so convoluted. (From Faith *et al.* (2001a). Reproduced with permission from Surrey Beatty & Sons Pty Ltd.)

not be conservation area candidates. Land-use intensity and a size of <10 km^2 were used as masks. In addition, any surrogate with an extent of <1 km^2 in any RMU was not counted as occurring in that RMU at all.

Land use intensity

RMUs that had more than half of their area in PNGRIS land use intensity classes from very high to moderate, a total of 954 RMUs, were masked out of the initial analysis. As it turned out, the biodiversity target could not be reached when this was done because some of these RMUs covered environments that were all subject to intensive land use. There were no unexploited examples of these environmental domains left in PNG. The strategy that was adopted in this case was to run the analysis with these RMUs masked out, get as close as possible to the target, save that set of RMUs and begin again with that set as the starting point, allowing the intensively used RMUs to be chosen. When this was done it was possible to reach the target with the addition of only 17 intensively used RMUs. These could be thought of as priorities for restoration or rehabilitation.

Small RMUs and small coverage by attributes

It was decided that RMUs $<10\,\mathrm{km}^2$ were too small to be candidate conservation areas because they would be unlikely to retain many of their species over the long term. This is an arbitrary cut-off and many areas larger than this are probably still too small. The most practical way to deal with this is to group small areas into larger management units and/or adopt sympathetic management outside the conservation areas.

When overlaying maps, for example environmental domains (straight line boundaries from clusters of grid cells), on RMUs (variable boundaries from air-photo patterns), small portions of domains may occur in RMUs to which they do not properly belong. In order not to count these as present in RMUs, any occurrence of domains or vegetation types of $<1\,\mathrm{km}^2$ was ignored in the analysis.

8.5.4 Preferences

Preferences refer to features that, all else being equal, it would be preferable to include in the conservation-area set. RMUs with low human population density and RMUs covering areas previously identified by experts as having high conservation value were the two preferences used for this conservation plan. The Conservation Needs Assessment (CNA) (Alcorn 1993; Beehler 1993) conducted some years prior to this study was used to incorporate expert opinion. It was achieved during the course of building up a priority-area set by maintaining masking and cost trade-offs operations, but looking first for a suitable RMU that was also a CNA priority area. Figure 8.11 shows RMUs representing areas identified as priorities by experts. Population density was extracted from PNGRIS and treated in the same way as expert opinion.

8.5.5 Priority conservation areas

Figure 8.12 shows the set of conservation areas selected to meet the biodiversity target, which also satisfies the objectives of minimizing opportunity costs, avoiding areas of high land-use intensity and high human population density, including existing protected areas and preferring CNA priority areas. This target was set by finding the number of environmental domains, when added to vegetation types that could be found in 10 % of the country. Table 8.3 summarizes this result. Altogether, 398 areas were needed to meet the goal. They covered 77 215 km², approximately 16.8 %

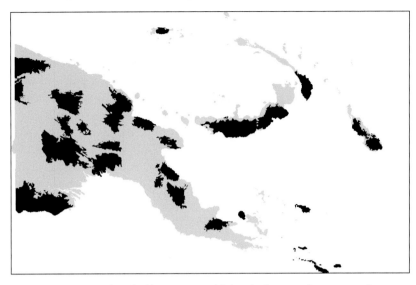

Figure 8.11. Areas identified by experts as high priority areas for conservation, part of the Conservation Needs Assessment (Beehler 1993). These areas have been adjusted to fit RMUs, which is why they are so convoluted. (From Faith *et al.* (2001a). Reproduced with permission from Surrey Beatty & Sons Pty Ltd.)

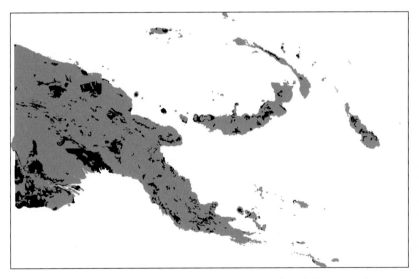

Figure 8.12. The 398 resource mapping units (RMUs) identified to meet the conservation goal and minimize foregone forestry opportunities, starting with existing protected areas, minimizing the number of areas with high land-use intensity and preferring areas with fewer people and areas previously identified by experts as having high conservation value, where there was a choice. (From Faith *et al.* (2001a). Reproduced with permission from Surrey Beatty & Sons Pty Ltd.)

Table 8.3 *Summary statistics for the selected set of conservation areas in Papua New Guinea. RMU refers to resource mapping unit, the planning units used in the study (Chapter 3, Example 3.1). CNA refers to the Conservation Needs Assessment (see text for details)*

Number of RMUs	398
Total area	77 215 km^2
Population	209 895
RMUs with agricultural potential	102
Timber volume index	93 218
RMUs from CNA priority areas	180
RMUs with high land-use intensity	17

of the country. Recall that in Chapter 4, Example 4.1, the number of environmental domains was determined by what could be sampled in 10% of PNG. Achieving that goal in the presence of land-cover change and opportunity costs required 16.8%. Only 102 of the selected areas had potential for agriculture and 180 coincided with expert opinion that their conservation value was high. A total of 209 895 people lived in them (at the time of the last census) out of a population of approximately four million.

This set of priority areas would inevitably be subject to review and re-evaluation as knowledge accumulates and as social and economic conditions change. The methods used were intended to function as an aid to conservation policy and decisions in an ongoing process. Data sets should be updated and analyses run again at regular intervals. Faith *et al.* (2001a,b) provide some examples of how this might be done. Currently (January 2006), priority areas for conservation action are being re-assessed throughout Melanesia by both Conservation International and WWF. Refinements of the methods described here are being used by both NGOs to aid this process.

8.6 Summary

In this chapter we summarized five cases in which the systematic methods described in this book have been used to produce conservation plans. We have not found a single case in which all the stages that we have described have been used, though many plans have incorporated most of the stages.

The Nullarbor example shows how a complementarity-based area selection can be carried out by hand without the use of computer-based algorithms. The Québec plan shows how distributions of species at risk can be used directly to formulate plans that have practical significance. It also shows how such analyses can be used to accommodate social concerns without jeopardizing biodiversity conservation. The California Channel Islands case shows how these methods, though mainly developed for terrestrial conservation, can straightforwardly be extended to the marine realm. The plan for the Cape Floristic Region is a case study in how to engage a wide range of collaborators and contributors. It pays particular attention to two important aspects of systematic conservation planning, the incorporation of ecological processes and how targets for the representation of surrogates can be explicitly determined. The Papua New Guinea study explicitly included a cost trade-off, comparing the biodiversity contribution of areas with their weighted cost, in order to reach the biodiversity target by minimizing forgone opportunities for forestry. Rightly or wrongly, this is important to planners and decision-makers in a country that derives much of its foreign exchange by logging its forests.

9 · *Conclusions*

In concluding this book, we want to reiterate strongly a statement made in the Introduction. The overall approach and the planning protocols described here are designed to help local experts make good policy decisions, not make policy on their behalf. They are decision support tools, not decision-making tools. It would be foolhardy to accept the results of a priority-setting exercise uncritically. This applies to all stages in the process. Data should be collected with their use in mind and knowing the likely method of data treatment. Collecting activities (called surveys in Chapter 3) should be designed carefully so that treatments (e.g., spatial models) are interpolations within the range of the data, not extrapolations beyond it. The mathematical and ecological assumptions of treatment methods should be understood and the methods applied critically. We must be prepared to re-examine the data if models seem unrealistic or if diagnostics indicate that models are poor fits to the data. Data treatment is a critical iterative process. So, too, is the application of area selection algorithms. As described in Chapters 5, 7 and 8, there can be many spatial arrangements of areas that will satisfy a biodiversity goal.

We should always be prepared to accept an alternative if it appears more likely to be implemented, even if it is not the optimum according to the data and area selection rules that have been applied. Similarly, we should not be overly concerned with getting computational algorithms exact or try too hard to find mathematically optimal solutions before using available methods to produce contextually relevant plans. Usually, the uncertain quality of the data and the vagaries of decision-making and policy formulation do not warrant such excessive theoretical efforts. We should always strive for improvement, but irreversible decisions on natural resource use are made every day. If we do not engage in the process, applying state-of-the-art science, the opportunity to do so will pass us by. The examples in Chapter 8 embody this attitude. They are all different and all are highly contextual, yet they are all grounded in good science and aimed squarely at supporting improved policy formulation and planning decisions to promote the protection of biodiversity.

We assume that the major users of systematic conservation planning tools will be government agencies responsible for nature conservation and natural resource management, and non-governmental organizations (NGOs) with conservation as a goal. We sincerely hope that industries, businesses and local communities will be active participants and that academic conservation biologists will continue to research and improve these planning tools. Sarkar *et al.* (2006) have recently reviewed the status of the concepts used to design planning tools described here. Some of the major investors of conservation resources in the world today are the big international NGOs, for example Conservation International (CI), The Nature Conservancy (TNC) and the World Wide Fund for Nature (WWF). Each of these NGOs focuses attention on relatively large regions (ecoregions, large watersheds, hotspots, conservation corridors) where they are dealing with many thousands of species, assemblages and environments. In order to plan conservation actions within these regions they need to prioritize areas. They have each adopted a systematic procedure akin to the methods described in this book. Both WWF and TNC use habitat types and environmental classes as well as selected species as biodiversity surrogates (Groves *et al.* 2002) and a staged planning process from data collection through goal setting to priority area identification. CI uses critically endangered species as true surrogates, maps the areas needed for these species to persist (key biodiversity areas; Eken *et al.* 2004) and then plans for the management of landscapes (called corridors) within which these key areas are located. We would also hope that, while these have sometimes been seen to be disparate approaches, they can now be seen, through this book, to be variations upon a common theme. All major conservation agencies are beginning to tread the same path.

9.1 Coping with uncertainty

Having said all that it must be acknowledged that there is still much uncertainty and a lack of precision throughout the conservation planning process. Our knowledge of biodiversity patterns and the processes that give rise to them can and should always be improved. New field surveys should target knowledge gaps and make most efficient use of the scarce resources available for conducting surveys, as described in Chapter 3. The data treatment and modeling techniques described in Chapter 4 also merit further development. The uncertainties embodied in the predictions from spatial models should always be explicitly recognized. Recent machine-learning methods that show promise, such as Maxent, which was described in Chapter 4, need to prove that they deliver accurate predictions repeatedly before they can displace other methods widely in use today.

Moreover, many of the techniques described in this book are yet at early stages of development and must be used with caution. Sarkar (2005) identified six questions relating to the scientific foundations of conservation biology that he believed merited closer philosophical attention. These same questions also merit closer scientific attention and four of them are relevant to systematic conservation planning:

(1) What if the problem of finding adequate estimator-surrogate sets has no solution? There are so few satisfactory tests of the adequacy of estimator-surrogate sets and so few cases in which adequate surrogates have been found that this is a very real possibility. Surrogates such as keystone and umbrella species, or taxa subsets such as birds, plants, etc. have not generally performed well when tested, though there have been occasional optimistic results. Environmental surrogates have also received equivocal reports. If adequate estimator surrogates cannot be found with current methods then there are two options: (a) Conservation plans will continue to be based on the surrogates in use today. It is just that they may not warrant description as surrogates for biodiversity. Protecting critically endangered species (the mission of Conservation International) and at least 10% of every major habitat type on Earth (the mission of The Nature Conservancy) will still contribute more to the protection of biodiversity than if these goals were not achieved. (b) Some methodology other than those currently in use may be devised to test the adequacy of estimator-surrogate sets. The more attention that is paid to this issue by the science community, the more likely it is that innovative methods will be developed. There is a clear need for closer scientific engagement with the problem of finding adequate estimator-surrogate sets.

(2) What if the viability problem cannot be solved? Ideally, systematic conservation planning requires the simultaneous quantitative extinction risk assessment of hundreds of species and other taxa in each area that is a candidate for inclusion in a conservation area network. It is quite possible that no adequate protocol can be developed towards this end; as Chapter 6 noted, population and even habitat-based viability analyses can presently only be performed for a few species at

a time. Nothing more ambitious seems immediately forthcoming. Unless we have a better handle on viability analysis the only options may be: (a) to perform viability analysis for as many species in each case as we can and hope for the best; (b) to use, as we often do in practice, expert intuition and experience to judge what constitutes a viable option for a habitat; or (c) to proceed by assuming, as we also often do in practice, that if we manage to maintain the integrity of a small identifiable set of ecological processes, biota in a habitat will remain secure.

(3) What if multiple criteria become almost always so incommensurable that acceptable compromises between conflicting interests can never be found? This is the scenario in which conservation becomes one of those intractable political problems with which we are familiar in many other areas of local, national and international policy. At present, globally, there is sufficient recognition of the importance of biodiversity (as indicated by the over 150 countries that have signed and ratified the Rio Convention on Biodiversity) that we do not yet face such a situation though there are many unresolved local conflicts. We must ensure that we minimize the chances of conflicting interests becoming unreconcilable by developing normative socially relevant rationales for the conservation of all components of biodiversity while explicitly acknowledging the legitimacy and relevance of many competing demands on natural resources, particularly for socio-economic and cultural development (Sarkar 2005). The goal must always be to develop consensus and community, not the unnecessary self-serving conflicts that conservation biologists have sometimes unfortunately encouraged (Guha 1989; Sarkar 1999).

(4) What if a satisfactory systematic method for incorporating uncertainty can never be worked out? In Chapter 6 we emphasized that planning for persistence is fraught with uncertainty and showed how some types of uncertainty may be incorporated into formal decision analysis, in particular, within a Bayesian framework. Though we did not discuss them in Chapter 7, protocols exist for multi-criteria analysis under uncertainty (see, for instance, Duenas & Mort 2002; Millet & Wedley 2002; Novikova & Pospelova 2002), even though these are yet to be used in systematic conservation plans such as those discussed in Chapter 8. However, the methods available at present may not provide sufficient guidance for decision-making in the presence of the numerous uncertainties associated with the conservation-planning process, for example: (a) uncertainties associated with data

collection and analysis for surrogacy assessment and place prioritization, (b) structural uncertainties associated with ecological models of viability analysis and (c) the lack of control over the uncertainties associated with the socio-political process that deals with incommensurate values. New methods of uncertainty analysis will almost certainly have to be devised and this is the domain of the philosophy of science. However, the important point to realize is that decisions must still be made. Given the rapid ongoing transformation of habitats throughout the world, uncertainty cannot be used as a pretext for not taking action.

The presence of such uncertainty merits two responses. The first, and most obvious, one is that we recognize the need for more research into addressing uncertainty at all stages of our analyses. This has now become an active area of research though we are far from reaching a consensus on how uncertainty should be modeled in conservation science (Regan et al. 2002). We have briefly discussed the use of a Bayesian framework (Chapter 6) but this is not the only option: proposals range from using fuzzy sets and imprecise probabilities, to models of information gaps (Regan et al. 2002). The second, and more important, response is to realize that no plan is final. Conservation plans must be updated periodically as more information flows in, not only because situations change in the field (that is, ecological and socio-economic features change over time) but because we have to revise the information on which a plan was based. Sometimes, uncertain knowledge may become precise, for instance, when we get better distributional data on biodiversity surrogates or have better models to predict biological and socio-economic features. We may also learn that what we regarded as precise, for instance, the habitat requirements of species, is actually uncertain. The requirement that plans must be routinely updated implies that systematic planning protocols be flexible, transparent and capable of rapid revision. We have emphasized these criteria throughout our discussion of the stages of systematic conservation planning.

9.2 Practicing conservation science in a complex world

We have explained the complexity of the concept of biodiversity, the consequent difficulty of operationalizing this concept for planning and policy making, and the need to resort to surrogates. Biodiversity is complex no matter how we look at it. Both the taxonomic and the spatial–ecological hierarchies are heterogeneous at all levels where there are untold viable configurations that can all be distinguished from one another. We are still a long way from an understanding of the ecosystem and evolutionary processes that give rise to biodiversity that is sound enough to guide management practices; so we must learn to live comfortably with uncertainty (Section 9.1). As if this amount of complexity is not enough, whenever we choose to participate in the conservation-planning process, we accept even more complexity in the form of the economic, socio-political and cultural contexts in which planning takes place.

Most people who come to conservation planning do so from a background in the biological or geographical sciences. As ecologists, for example, we might understand population dynamics, hydrological processes and predator–prey relationships, and we can probably model them successfully. We do these things almost routinely and, we hope that we are good at them. Lately, we are being asked to go further by weighing compromises between production and protection, understanding the institutional and governance arrangements that drive natural resource management and decision making and incorporating multiple criteria into future landscape scenarios. When we do this we dabble in fields that rightly belong to other disciplines and we feel uncomfortable. But if we do not take a lead, no one will. Decisions on the use of natural resources will continue to be made mainly on narrow short-term economic and social grounds. If on occasion environmental concerns are taken on account, these will mostly be limited to negative environmental impacts on production or the generation of incomes. Institutional mechanisms and governance arrangements will continue to favor biodiversity depletion over conservation and sustainable development. It is primarily scientists, biologists, ecologists, geographers, etc., and especially conservation biologists, who have both the knowledge of the need, and the will, to bring the biodiversity component of the natural resource base into the planning and policy arena.

The good news is that we do not have to become experts in all these fields ourselves because we can form partnerships and collaborate with our peers in other fields. What we must do is fully accept the need for this collaboration and build enough understanding of other disciplines so that we know how to work with them productively. We must learn to speak their languages. We need to understand how variables from all sectors interact and to work at the intersections of disciplinary understandings. There is one more step we need to take. We need to embed our science in regions, working with the people who influence and use natural resources, not be wheeled in to solve technical problems and then wheeled out again. The plan developed in the Cape Floristic Region (Chapters 6 and 8) is a good example.

This leads on to the importance of context and a reiteration of the observation that in future the most valuable science, especially biological science, might well consist of building highly contextual procedures rather than producing or seeking general theories (Sarkar 2004). The process of building conservation plans is as much a product as the plan itself. This is especially true since conservation planning must be scheduled over a period of years. As knowledge accumulates, social and economic conditions change and institutions evolve, conservation priorities will likely change so plans will always be dynamic evolving things.

9.3 Future directions

This book is just a glimpse of a future multi-faceted interdisciplinary field that includes conservation planning as an integral part of sustainable landscape planning and management. We use the term "landscape" here for ease of exposition. We mean to include seascapes in the scope of our discussion and we believe that seascapes require the same approach. The entire conservation planning process is a research agenda for the future. There are important unresolved issues at every stage and we need to create new knowledge and a better understanding of how to do systematic conservation planning. This includes improving the quality and precision of data, a resolution of the surrogacy issue, if possible or, if not, at least a consensus on appropriate measures of biodiversity and the iterative improvement of decision support tools. But the greatest challenges are embedded in the very first step of the framework described in Chapter 1 (Section 1.3): we must fully integrate biodiversity conservation into the policy, planning and management processes that drive the use of natural resources. The bottom line here is that the production of livelihoods and the creation of profits must be integrated with biodiversity protection because without that integration, the fate of biodiversity is necessarily bleak. Chapter 7 explored some of the methods available for helping to do this.

In the future, as we develop frameworks beyond what we have discussed here, and embed the planning protocol for biodiversity within a broader vision of how we live, we must address sustainability explicitly (Norton 2003). We must also accept the need for, and the cost of, environmental restoration – or reconstruction (Perrow & Davy 2002) – to attempt to reverse some of the harm that past over-consumption has done to landscapes. In our view, ongoing intervention is a necessary component of adequate management: what remains to be decided is whether intervention is patchy, insular, guided by the disparate goals of different interest groups, or integrated to achieve some level of agreement between interest groups. We suggest the latter option, however difficult it may initially seem to attain such agreement. Here we must expand beyond the natural and social sciences and embrace the humanities, learning from environmental historians and philosophers how humans envision their futures and act to achieve those visions. We must understand how people generate and respond to values, and how those values change.

Even as we begin such a process, we should acknowledge that it must be underpinned by sound information on the biological and physical components of the landscape. The protocols and methods described in Chapters 2, 3 and 4 are essential for two purposes. One is to derive the biological and environmental data needed as input to the planning process, that is, deciding exactly what biodiversity surrogates we want to protect and mapping them. The other is as a baseline against which future success or failure can be judged. This need is so obvious, yet such a baseline is not widely available at present. Investments in systematic structured data collection and data treatment activities are few and far between. More money and effort need to be invested to compile this baseline. We also need to acknowledge the continuum from the more or less pristine environments of large national parks and world heritage areas, through fragmented and variegated landscapes, to landscapes that have been largely transformed through human activities by, for example, cultivation and settlements. We need to accept that we cannot protect all of the biodiversity along this continuum. Some, perhaps substantial, loss is inevitable. But what we can do is work to minimize that loss where it is continuing, restore some of the biodiversity in transformed landscapes, and ensure the persistence of biodiversity in conservation areas. Systematic conservation planning is a key component of designing sustainable landscapes, but it is not the only one. In fragmented and transformed landscapes we also need to manage ecosystem functions to ensure that they persist and we need to develop policies and economic processes that will assist biodiversity and ecosystem processes to help pay for themselves.

Reducing the impacts of land use change on biodiversity has been one of the most intractable problems for conservation biologists and managers. This is at least partly because biodiversity management programs have focused on defining species-specific management plans and habitat quality standards only in ecological terms. Generally, this has so far been unsuccessful as there is still no rigorous or widely accepted ecological approach for determining the optimal suite of actions to meet those standards. In addition, identifying threatening processes and their impacts, understanding the role of institutions and policy mechanisms (and learning how to change these), quantifying the cost and efficacy of management alternatives and creating truly adaptive implementable management programs are all activities that are required for biodiversity conservation to succeed. Better strategies (and there will be no generally optimal strategy, just better or worse ones) will depend on the impacts of land uses, the range of existing and potential management practices, and their cost-effectiveness across the landscape. Implementation must be spatially explicit and will require that natural-resource ownership and property rights issues are fully acknowledged and taken into account, that costs are acceptable to a wide range of stakeholders and that institutional arrangements support implementation.

9.3.1 Agricultural landscapes

Agriculture was one of the earliest forms of land transformation and it is continuing to have a major impact today, especially as large agribusinesses adopt new technologies for intensive production. The trend continues, for example, toward monocultures which are easy to harvest and intensive feed lots for the production of meat. As noted above, biodiversity protection will never win out over the creation of wealth in the short to medium term. But in the long term, a loss of biodiversity will reduce options for wealth creation and ultimately lead to a loss of income, particularly in the farming and tourism sectors. It will also lead to expensive repair and restoration programs. In parts of Western Australia, it has been estimated that by 2050, due to salinity, $400 m will be lost in agricultural production and sealed road life expectancy will be reduced by 75 % (Agriculture WA; www.agric.wa.gov.au/). So, perhaps, it would make sense to buck the trend and work to develop farming systems that are both profitable and environmentally sustainable?

Innovation in farming systems to create multiple benefits is emerging as a key strategic regional research and development activity. This activity aims to achieve improved financial returns for industry while contributing solutions to environmental degradation and creating social benefits for communities. Internationally, there is growing momentum behind the concept of "eco-agriculture," which aims to integrate innovation in farming systems into the management of landscapes to sustain livelihoods, biodiversity and ecosystem services (McNeely & Scherr 2003). We will not attempt to deal with these issues in any depth. Suffice it to say that economic models of diversified farming systems suggest that they are viable in most places over the long term so long as they are backed by appropriate infrastructure development, and appropriate policy support in the form of, for example, market-based instruments, institutional incentives and environmental credits (Abadi *et al.* 2003), as well as a market development strategy and sound economic planning. There is a clear and present need for policy-makers to take the initiative here.

Another option, at least in those many parts of the world that are still being transformed, is to reverse the standard development paradigm (Nix 2004). Instead of allowing development to proceed piecemeal until degradation is severe, why not begin with a whole of landscape assessment that identifies the level of development that is sustainable for both livelihoods and the environment? This development cap need not remain fixed, but could be modified with new research findings and monitoring. Social and economic factors would play a major role in determining an acceptable development cap, but so too would our knowledge of biodiversity and ecosystem processes.

Adoption of best management practices by farmers has often been identified as a key step to improving ecosystem function and protecting biodiversity. This is more likely to happen when both incentives for adoption are applied and impediments to their implementation are removed. This requires an understanding of the attitudes, values and beliefs of resource owners and their perceptions of potential trade-offs and impediments to adoption of best management practices. Thus, an important research need is the development of appropriate methods to assess and determine key impediments, incentives and trade-offs for different management strategies.

9.3.2 Urbanizing landscapes

Urbanization is the most recent form of land transformation and it is becoming ubiquitous (McKinney 2002; Huston 2005). Human populations are growing and settlement is expanding, impacting on productive agricultural land and biodiversity. In some cases the reason is simply pressure from increasing population size. In others, lifestyle choices mean people are moving to small acreages on the fringes of towns and cities, gradually "urbanizing" the countryside. In yet others, infrastructure development, especially roads, for example in the Amazon, is opening up opportunities for settlement. A challenge is to manage these landscapes so that biodiversity, livelihoods and lifestyles are all sustained. Studying such landscapes in transition provides a start to developing management that addresses sustainability explicitly. It is also practical because there is already some interest in developing and implementing systematic plans for such regions.

To model landscapes tending towards urbanization, knowledge on bio-diversity has to be collected and the services provided by ecosystem processes must be understood. These include, for example, the regulation of disturbances such as floods and fires, erosion control, the regulation of acidity and salinity in soils and water, the regulation of nutrients and oxygen levels in water and land suitability for recreation, scenic amenity and cultural values. Using this knowledge base, a research agenda might include the following objectives:

• Develop plausible scenarios of biodiversity, land and water management for landscapes in transition;
• Identify novel combinations of productive land and water and their management across these landscapes;
• Design, experiment and model biodiversity and productive land and water typologies in these landscapes;
• Develop techniques to assist people make more informed choices about biodiversity, land and water management;
• Articulate "preferred" scenarios through planning and governance mechanisms;
• Embed new forms of landscape management through capacity building.

In summary, what we need to do is combine knowledge of biodiversity patterns and processes with a social, economic and cultural understanding of human communities, an appreciation of how people imagine their futures to be and how they act to achieve those futures, and an understanding of the institutional and governance arrangements for the use of natural resources, in order to explore how landscapes might evolve and the likely impacts of different change pathways on biodiversity. This could readily be extended beyond biodiversity to include ecosystem processes such as stocks and flows of water, nutrients and energy and the dispersal of propagules. The key outputs would be planning tools built from models of socio-economic processes of change and their environmental consequences. Such tools would address the dynamic processes leading to change in landscapes and identify the control points available to policy makers and regional planners for steering the direction of change, and balancing desirable socio-economic goals with the maintenance of biodiversity.

9.4 Summary

The planning protocols described in this book are decision-support tools, not decision-making tools. They should be used critically, with all stages in the planning process (Table 1.1) subject to revision as applications reveal new problems and issues, and research proceeds to solve those problems. To some extent this is already happening as both international conservation NGOs and natural resource management agencies within countries apply more systematic methods to conservation planning, but this process of iterative improvement needs to become embedded in the cultures of both research providers and research users. Too often providers deliver elegant solutions to problems that cannot find traction on the ground, in which case users turn away from potentially valuable new tools, possibly entrenching the mistakes of the past, which led to unrepresentative and insecure conservation area networks in the first place.

Uncertainty constrains the application of these tools at all stages of the planning process from data collection and treatment right through to implementation and revision. More research will help reduce uncertainty, but it will never disappear. Accepting that plans are dynamic and should adapt to new knowledge and changing social and economic conditions is the way to live with uncertainty; planning is an ongoing iterative process. Uncertainty is related to complexity. Biological complexity limits our understanding of what management prescriptions are needed to sustain biodiversity. The complexity of the social, economic and cultural contexts that societies operate in limits our ability to achieve real biodiversity conservation outcomes. Conservation biologists need to embrace the social sciences and humanities because biodiversity conservation will not be achieved without economic benefit, or without social and cultural change. This is the new challenge for conservation biology; to fully integrate biodiversity conservation into the policy, planning and management processes that govern the use of natural resources.

References

Abadi, A., Lefroy, T., Cooper, D., Hean, R. & Davies, C. (2003). *Profitability of Medium to Low Rainfall Agroforestry in the Cropping Zone*. Final Report, RIRDC Project No. UWA63A. Rural Industries R&D Corporation, Canberra, Australia, p. 63.

Ackery, P. R. & Vane-Wright, R. I. (1984). *Milkweed Butterflies*, Cornell: Cornell University Press.

Airamé, S., Dugan, J. E., Lafferty, K. D. *et al.* (2003). Applying ecological criteria to marine reserve design: a case study from the California Channel Islands. *Ecological Applications*, **13**, S170–84.

Akçakaya, H. R. (2000). Viability analysis with habitat-based metapopulation models. *Population Ecology*, **42**, 45–53.

Akçakaya, H. R. & Atwood, J. L. (1997). A habitat-based metapopulation model of the California gnatcatcher. *Conservation Biology*, **11**, 422–434.

Alcorn, J. B. (ed.) (1993). *Papua New Guinea Conservation Needs Assessment Vol. 1. Biodiversity Support Program*, Landover, Maryland: Government of Papua New Guinea, Department of Environment & Conservation, Corporate Press Inc.

Alford, A. (2005). Wildlife tax breaks lack state oversight. *The Austin-American Statesman*. Austin, Texas, May 29.

Alldredge, J. R. & Ratti, J. T. (1992). Further comparison of some statistical techniques for analysis of resource selection. *Journal of Wildlife Management*, **56**, 1–9.

Ananda, J. & Herath, G. (2003). The use of the analytic hierarchy process to incorporate stakeholder preferences into regional forest planning. *Forest Policy and Economics*, **5**, 13–26.

Andelman, S. J., Ball, I., Davis, F. & Stoms, D. (1999). *Sites V 1.0. An Analytical Toolbox for Designing Ecoregional Conservation Portfolios*, Santa Barbara: National Center for Ecological Analysis & Synthesis.

Andelman, S. J. & Fagan, W. F. (2000). Umbrellas and flagships: efficient conservation surrogates, or expensive mistakes? *Proceedings of the National Academy of Sciences (USA)*, **97**, 5954–9.

Anderberg, M. R. (1973). *Cluster Analysis for Applications*, New York: Academic Press.

Anderson, R. P. (2003). Real vs. artefactual absences in species distributions: tests for *Oryzomys albigularis* (Rodentia: Muridae) in Venezuela. *Journal of Biogeography*, **30**, 591–605.

Anderson, R. P., Gómez-Laverde, M. & Peterson, A. T. (2002). Geographical distributions of spiny pocket mice in South America: insights from predictive models. *Global Ecology and Biogeography*, **11**, 131–41.

Anderson, R. P. & Martínez-Meyer, E. (2004). Modeling species' geographic distributions for preliminary conservation assessments: an implementation with the spiny pocket mice (*Heteromys*) of Ecuador. *Biological Conservation*, **116**, 167–79.

Anselin, A., Miere, P. & Anselin, M. (1989). Multicriteria techniques in ecological evaluation: an example using the Analytic Hierarchy Process. *Biological Conservation*, **49**, 215–29.

Araújo, M. B., Densham, P. J. & Humphries, C. (2003). Predicting species diversity with ED: the quest for evidence. *Ecography*, **26**, 380–3.

Araújo, M. B., Densham, P. J., Lampinen, R. *et al.* (2001). Would environmental diversity be a good surrogate for species diversity? *Ecography* **24**: 103–10.

Araújo, M. B., Densham, P. J. & Williams, P. H. (2004). Representing species in reserves from patterns of assemblage diversity. *J. Biogeography* **31**, 1037–50.

Arrow, K. J. & Raynaud, H. (1986). *Social Choice and Multicriterion Decision-Making*, Cambridge, MA: The MIT Press.

Austin, M. P. (1985). Continuum concept, ordination methods and niche theory. *Annual Review of Ecology and Systematics*, **16**, 39–61.

Austin, M. P. (1987). Models for the analysis of species response to environmental gradients. *Vegetatio*, **69**, 35–45.

Austin, M. P. (2002). Spatial prediction of species distribution: an interface between ecological theory and statistical modelling. *Ecological Modelling*, **157**, 101–18.

Austin, M. P. (2005). Vegetation and environment: discontinuities and continuities. In *Vegetation Ecology*, ed. E. van der Maarel. Oxford: Blackwell Publishing, pp. 52–84.

Austin, M. P. & Belbin, L. (1982). A new approach to the species classification problem in floristic analysis. *Australian Journal of Ecology*, **7**, 75–89.

Austin, M. P., Cawsey, E. M., Baker, B. L. *et al.* (2000). *Predicted Vegetation Cover in the Central Lachlan Region. Final Report of the Natural Heritage Trust Project AA1368.97.* Canberra: CSIRO Wildlife & Ecology.

Austin, M. P., Cunningham, R. B. & Fleming, P. M. (1984). New approaches to direct gradient analysis using environmental scalars and statistical curve-fitting procedures. *Vegetatio*, **55**, 11–27.

Austin, M. P. & Heyligers, P. C. (1989). Vegetation survey design for conservation: gradsect sampling of forests in north-eastern New South Wales. *Biological Conservation*, **50**, 13–32.

Austin, M. P. & Heyligers, P. C. (1991). New approach to vegetation survey design: gradsect sampling. In *Nature Conservation: Cost Effective Biological Surveys and Data Analysis*, ed. C. R. Margules & M. P. Austin. Melbourne: CSIRO, 31–41.

Austin, M. P. & Margules, C. R. (1986). Assessing representativeness. In *Wildlife Conservation Evaluation*, ed. M. B. Usher. London: Chapman & Hall, pp. 47–52.

Austin, M. P. & Meyers, J. A. (1996). Current approaches to modelling the environmental niche of eucalypts: implications for management of forest biodiversity. *Forest Ecology and Management*, **85**, 95–106.

Austin, M. P., Nicholls, A. O., Doherty, M. D. & Meyers, J. A. (1994). Determining species response functions to an environmental gradient by means of a β-function. *Journal of Vegetation Science*, **5**, 215–28.

Austin, M. P., Nicholls, A. O. & Margules, C. R. (1990). Measurement of the realized qualitative niche: environmental niches of five *Eucalyptus* species. *Ecological Monographs*, **60**, 161–77.

Austin, M. P. & Smith, T. M. (1989). A new model for the continuum concept. *Vegetatio*, **83**, 35–47.

Ball I. R. & Possingham, H. P. (2000). MARXAN version (1.8.2) User's Manual www.ecology.uq.edu.au/marxan.htm.

Bastedo, J. D. (1986). *An ABC Resource Survey Method for Environmentally Significant Areas with Special Reference to Biotic Surveys in Canada's North.* Dept. of Geography Publications Series No. 24. University of Waterloo, Canada.

Bateman, I. J. & Willis, K. G. (eds.) (1999). *Valuing Environmental Preferences.* Oxford: Oxford University Press.

Beard, J. S. (1975). *The Vegetation of the Nullarbor Area. Vegetation Survey of Western Australia, 1:1,000,000 Series, Explanatory Notes to Sheet 4, Nullarbor.* Perth: University of Western Australia Press.

Beehler, B. M. (ed.) (1993). *Papua New Guinea Conservation Needs Assessment Vol. 2. Biodiversity Support Program*, Landover, Maryland: Corporate Press Inc & Government of Papua New Guinea, Department of Environment & Conservation.

Beissinger, S. (2002). Population viability analysis: past, present, future. In *Population Viability Analysis*, ed. S. Beissinger & D. R. McCullough. Chicago: University of Chicago Press, pp. 5–17.

Beissinger, S. & McCullough, D. R. (eds.) (2002). *Population Viability Analysis.* Chicago: University of Chicago Press.

Belbin, L. (1980). *TWOSTEP: a program incorporating asymmetric comparisons that uses two steps to produce a dissimilarity matrix. CSIRO Division of Land Use Research Technical Memorandum 80/9*, Canberra: CSIRO.

Belbin, L. (1984). Fuse: a FORTRAN 4 program for agglomerative fusion for mini-computers. *Computers & Geosciences*, **10**, 361–84.

Belbin, L. (1987). The use of non-hierarchical allocation methods for clustering large sets of data. *Australian Computer Journal*, **19**, 32–41.

Belbin, L. (1991). The analysis of pattern in bio-survey data. In *Nature Conservation: Cost-effective Surveys and Data Analysis*, ed. C. R. Margules & M. P. Austin. Melbourne: CSIRO, pp. 176–90.

Belbin, L. (1993). *PATN Users Guide*, Canberra: CSIRO Division of Wildlife & Ecology.

Belbin, L., Austin, M. P., Margules, C. R., Cresswell, I. D. & Thackway, R. (1994). *Data Suitability: Sub-project 1, Modelling of Landscape Patterns and Processes Using Biological Data.* Canberra: Division of Wildlife & Ecology, Commonwealth Scientific and Industrial Research Organisation.

Bellamy, J. A. & McAlpine, J. R. (1995). *Papua New Guinea: Inventory of Natural Resources, Population Distribution and Land Use Handbook*, 2nd edn. PNGRIS Publication No. 6. Canberra: AusAID.

Belton, V. (1986). A comparison of the analytic hierarchy process and a simple multi-attribute value function. *European Journal of Operational Research*, **26**, 7–21.

Belton, V. & Gear, T. (1983). On a short-coming of Saaty's method of analytic hierarchies. *Omega*, **11**, 226–30.

Bengtsson, J., Angelstam, P., Elmqvist, T. *et al.* (2003). Reserves, resilience and dynamic landscapes. *Ambio*, **32**, 389–96.

Berbel, J. & Zamora, R. (1995). An application of MOP and GP to wildlife management (deer). *Journal of Environmental Management*, **44**, 29–38.

Bogetoft, P. & Pruzan, P. (1997). *Planning with Multiple Criteria: Investigation, Communication and Choice*, Copenhagen: Copenhagen Business School Press.

Bojórquez-Tapia, L., de la Cueva, H., Díaz, S. *et al.* (2004). Environmental conflicts and nature reserves: redesigning Sierra San Pedro Mártir National Park, Mexico. *Biological Conservation*, **117**, 111–26.

Borrini-Feyerabend, G. (1996). *Collaborative Management of Protected Areas: Tailoring the Approach to the Context. Issues in Social Policy.* Gland, Switzerland: IUCN.

Bowne, D. R. & Bowers, M. A. (2004). Interpatch movements in spatially structured populations: a literature review. *Landscape Ecology*, **19**, 1–20.

Boyce, M. S. (1992). Population viability analysis. *Annual Review of Ecology and Systematics*, **23**, 481–506.

Boyce, M. S., Meyer, J. S. & Irwin, L. (1994). Habitat-Based PVA for the Northern Spotted Owl. In *Statistics in Ecology and Environmental Monitoring*, ed. D. J. Fletcher & B. F. J. Manly. Dunedin, N.Z.: University of Otago Press, pp. 63–85.

Braithwaite, L. W., Austin, M. P., Clayton, M., Turner, J. & Nicholls, A. (1989). On predicting the presence of birds in *Eucalyptus* forest types. *Biological Conservation*, **50**, 33–50.

Braithwaite, L. W., Dudzinski, M. L. & Turner, J. (1983). Studies of the arboreal marsupial fauna of eucalypt forests being harvested for wood pulp at Eden, New South Wales. II. Relationship between the fauna density, richness and diversity and measured variables of habitat. *Australian Wildlife Research*, **10**, 231–47.

Braithwaite, L. W., Turner, J. & Kelly, J. (1984). Studies of the arboreal marsupial fauna of eucalypt forests being harvested for wood pulp at Eden, New South Wales. III. Relationships between fauna densities, eucalypt occurrence and foliage nutrients and soil parent materials. *Australian Wildlife Research*, **11**, 41–8.

Branch, W. R. (1988). *South African Red Data Book – Reptiles and Amphibians. South African National Programmes Report No 151.* Pretoria: CSIR.

Bray, J. R. & Curtis J. T. (1957). An ordination of upland forest communities of southern Wisconsin. *Ecological Monographs*, **27**, 325–49.

Brook, B. W., Chapman, A. P., Burgman, M. A., Akcakaya, H. R. & Frankham, R. (2000). Predictive accuracy of population viability analysis in conservation biology. *Nature*, **404**, 385–7.

Brower, L. P. & Malcolm, S. B. (1991). Animal migrations: endangered phenomena. *American Zoologist*, **31**, 265–76.

Bull, A. L., Thackway, R. & Cresswell, I. D. (1993). *Assessing Conservation of the Major Murray-Darling Ecosystems*. Canberra: Environmental Resources Information Network (ERIN), Australian Nature Conservation Agency.

Burbidge, A. H., Harvey, M. S. & McKenzie, N. L. (eds.) (2000). Biodiversity of the southern Carnarvon Basin. *Records of the Western Australian Museum*, Supplement No. 61, Perth.

Burgman, M. A., Ferson, S. & Akçakaya, H. R. (1993). *Risk Assessment in Conservation Biology*, London: Chapman & Hall.

Burgman, M. A., Possingham, H. P., Lynch, A. J. J. *et al.* (2001). A method for setting the size of plant conservation target areas. *Conservation Biology*, **15**, 603–16.

Burnham, K. P. & Anderson, D. R. (1998). *Model Selection and Inference*. New York: Springer-Verlag.

Burrough, P. A. (1986). *Principles of Geographical Information Systems for Land Resources Assessment*. New York: Oxford University Press.

Burton, J. A. (1984). A bibliography of Red Data books. *Oryx*, **18**, 61–4.

Busby, J. R. (1986). A biogeoclimatic analysis of *Nothofagus cunninghamii* (Hook) Oerst in eastern Australia. *Australian Journal of Ecology*, **11**, 1–7.

Busby, J. R. (1991). BIOCLIM – a bioclimatic analysis and prediction system. In *Nature Conservation: Cost Effective Biological Surveys and Data Analysis*, ed. C. R. Margules & M. P. Austin. Melbourne: CSIRO, pp. 64–8.

Butler, J. C., Jia, J. & Dyer, J. S. (1997). Simulation techniques for the sensitivity analysis of multi-criteria decision models. *European Journal of Operational Research*, **103**, 531–45.

Cabeza, M. & Moilanen, A. (2001). Design of reserve networks and the persistence of biodiversity. *Trends in Ecology and Evolution*, **5**, 242–8.

Camm, J. D., Norman, S. K., Polasky, S. & Solow, A. (2002). Nature reserve site selection to maximize expected species covered. *Operations Research*, **50**, 946–55.

Cantú, C., Wright, R. G., Scott, J. M. & Strand, E. (2004). Assessment of current and proposed nature reserves of Mexico based on their capacity to protect geophysical features and biodiversity. *Biological Conservation*, **115**, 411–17.

Caro, T. M. & O'Doherty, G. (1999). On the use of surrogate species in conservation biology. *Conservation Biology*, **13**, 805–14.

Carpenter, G., Gillison, A. N. & Winter, J. (1993). DOMAIN: a flexible modelling procedure for mapping potential distributions of plants and animals. *Biodiversity and Conservation*, **2**, 667–80.

Caughley, G. (1994). Directions in conservation biology. *Journal of Animal Ecology*, **63**, 215–44.

Caughley, G. C. & Gunn, A. (1996). *Conservation Biology in Theory and Practice*. Cambridge, Massachusetts: Blackwell Science.

Cawsey, E. M., Austin, M. P. & Baker, B. L. (2002). Regional vegetation mapping in Australia: a case study in the practical use of statistical modelling. *Biodiversity and Conservation*, **11**, 2239–74.

Chomitz, K. M., Brenes, E. & Constantino, L. (1999). Financing environmental services: the Costa Rican experience and its implications. *The Science of the Total Environment*, **240**, 157–69.

Christian, C. S. & Stewart, G. A. (1968). Methodology of integrated surveys. In *Proceedings of the Toulouse Conference on Aerial Surveys and Integrated Studies*. Paris: UNESCO, pp. 233–80.

Church, R. L., Stoms, D. M. & Davis, F. W. (1996). Reserve selection as maximal covering location problem. *Biological Conservation*, **76**, 105–12.

Clevenger, A. P., Wierzchowski, J., Chruszcz, B. & Gunson, K. (2002). GIS-generated, expert-based models for identifying wildlife habitat linkages and planning mitigation passages. *Conservation Biology*, **16**(2), 503–14.

Cocks, K. D. & Baird, I. A. (1989). Using mathematical programming to address the multiple reserve selection problem: an example from the Eyre Peninsula, South Australia. *Biological Conservation*, **49**, 113–30.

Cohon, J. L. (1978). *Multiobjective Programming and Planning*. New York: Academic Press.

Colwell, R. K. & Coddington, J. A. (1994). Estimating terrestrial biodiversity through extrapolation. *Philosophical Transactions of the Royal Society of London Series B*, **345**, 101–18.

Commonwealth of Australia (1997). *Nationally Agreed Criteria for the Establishment of a Comprehensive, Adequate and Representative Reserve System for Forests in Australia*. Canberra: Australian Government Publishing Service

Conservation International (2004). *Conserving Earth's Living Heritage: a Proposed Framework for Designing Biodiversity Conservation Strategies*. Washington, D.C.: Conservation International.

Cowling, R. M. (ed.). (1992). *The Ecology of Fynbos*, Cape Town: Oxford University Press.

Cowling, R. M. & Heijnis, C. E. (2001). The identification of broad habitat units as biodiversity entities for systematic conservation planning in the Cape Floristic Region. *South African Journal of Botany*, **67**, 15–38.

Cowling, R. M. & Pressey, R. L. (2003). Introduction to systematic conservation planning in the Cape Floristic Region. *Biological Conservation*, **112**, 1–13.

Cowling, R. M., Pressey, R. L., Rouget, M. & Lombard, A. T. (2003). A conservation plan for a global biodiversity hotspot – the Cape Floristic Region, South Africa. *Biological Conservation*, **112**, 191–216.

Csuti, B., Polasky, S., Williams, P. H. *et al.* (1997). A comparison of reserve selection algorithms using data on terrestrial vertebrates in Oregon. *Biological Conservation*, **80**, 83–97.

Curry, G. (1987). Geology and geomorphology. In *A Biological Survey of the Nullarbor Region South and Western Australia in 1984*, ed. N. L. McKenzie & A. C. Robinson. Adelaide: South Australian Department of Environment and Planning, pp. 17–22.

Cutter, S. L. (1996). Vulnerability to environmental hazards. *Progress in Human Geography*, **20**, 529–39.

Dennis, B. (1996). Discussion: should ecologists become Bayesians? *Ecological Applications*, **6**, 1095–103.

Diamond, J. M. (1975). The island dilemma: lessons of modern biogeographic studies for the design of natural reserves. *Biological Conservation* **7**, 129–46.

Diamond, J. M. & May, R. M. (1976). Island biogeography and the design of natural reserves. In *Theoretical Ecology: Principles and Applications*, ed. R. M. May. Oxford: Blackwell, pp. 163–86.

Dias, P. (1996). Sources and sinks in population biology. *Trends in Ecology and Evolution*, **11**, 326–30.

Dickman, C. R., Pressey, R. L., Lim, L. & Parnaby, H. E. (1993). Mammals of particular conservation concern in the Western Division of New South Wales. *Biological Conservation*, **65**, 219–48.

Dilley, M. & Boudreau, T. E. (2001). Coming to terms with vulnerability: a critique of the Food Security definition. *Food Policy*, **26**, 229–47.

Dinerstein, E., Olson, D. M., Graham, D. J. *et al.* (1995). Ecoregions of Latin America and the Caribbean. In *A Conservation Assessment of the Terrestrial Ecoregions of Latin America and the Caribbean*. Washington, DC: The World Bank.

Disney, R. H. L. (1986). Inventory surveys of insect faunas: discussion of a particular attempt. *Antenna*, **10**, 112–16.

Dobson, A. P., Rodriguez, J. P., Roberts, W. M. & Wilcove, D. S. (1997). Geographical distribution of endangered species in the United States. *Science*, **275**, 550–3.

Dowie, M. (2005). Conservation refugees. *Orion*, November–December, 16–27.

Duenas, A. & Mort, N. (2002). Solving a multiple criteria decision-making problem under uncertainty. *Intelligent Systems, 2002 Proceedings. First International IEEE Symposium*, **1**, 308–13.

Durant, S. M. & Mace, G. M. (1994). Species differences and population structure in population viability analysis. In *Creative Conservation: Interactive Management of Wild and Captive Animals*, ed. P. J. S. Olney, G. M. Mace & A. T. C. Feistner. London: Chapman & Hall, pp. 67–91.

Dyer, J. (1990). Remarks on the analytic hierarchy process. *Management Science*, **36**, 249–2.

Dyer, J. S. (2005). MAUT – Multi attribute utility theory. In *Multiple Criteria Decision Analysis: State of the Art Surveys*, ed. J. Figueira, S. Greco & M. Ehrgott. Dordrecht: Kluwer, pp. 265–94.

Dyer, J., Fishburn, P., Steuer, R., Wallenius, J. & Zionts, S. (1992). Multiple criteria decision making, multi attribute utility theory: the next ten years. *Management Science*, **38**, 645–54.

East, R. (1988). Summary of regional status of antelopes in east and north-east Africa. In *Antelopes: Global Survey and Regional Action Plans, Part 1.*, ed. R. East. Gland: IUCN, pp. 71–6.

East, R. (1989). Summary of regional status of antelopes in southern and south-central Africa. In *Antelopes: Global Survey and Regional Action Plans, Part 2.*, ed. R. East. Gland: IUCN, pp. 76–79.

Eken, G., Bennon, L., Brooks, T. M. *et al.* (2004). Key biodiversity areas as site conservation targets. *Bioscience*, **54**, 1110–18.

Elith, J. & Burgman, M. A. (2002). Predictions and their validation: rare plants in the Central Highlands, Victoria, Australia. In *Predicting Species Occurrences: Issues of Accuracy and Scale*, ed. J. M. Scott, P. J. Heglund, M. L. Morrison, M. G. Raphael, W. A. Wall & F. B. Samson. Covelo, CA: Island Press, pp. 303–314.

Elith, J., Graham, C. H., Anderson, R. P. *et al.* (2006). Novel methods improve prediction of species' distributions from occurrence data. *Ecography*, **29**, 129–51.

Ellison, A. M. (1996). An introduction to Bayesian inference for ecological research and environmental decision-making. *Ecological Applications*, **6**, 1036–46.

Environmental Protection Agency (2002). *Biodiversity Assessment and Mapping Methodology, Version 2.1 July 2002*. Biodiversity Branch, Environmental Protection Agency, Queensland, Australia. www.epa.qld.gov.au/publications/p00471aa. pdf/ Biodiversity_assessment_and_mapping_methodology.pdf

Erlich, P. R. & Wilson, E. O. (1991). Biodiversity studies: science and policy. *Science*, **253**, 758–62.

Erwin, T. L. (1982). Tropical forests: their richness in Coleoptera and other arthropod species. *Coleopterists' Bulletin*, **36**, 74–5.

Erwin, T. L. (1983). Tropical forest canopies: the last biotic frontier. *Bulletin of the Entomological Society of America*, **30**, 14–19.

Faith, D. P. (1991). Effective pattern analysis methods for nature conservation. In *Nature Conservation: Cost-effective Surveys and Data Analysis*, ed. C. R. Margules & M. P. Austin. Melbourne: CSIRO, pp. 47–53.

Faith, D. P. (1992). Conservation evaluation and phylogenetic diversity. *Biological Conservation*, **61**, 1–10.

Faith, D. P. (1994). Phylogenetic pattern and the quantification of organismal diversity. *Philosophical Transactions of the Royal Society of London Series B*, **345**, 45–58.

Faith, D. P. (1995). *Biodiversity and Regional Sustainability Analysis*. Lyneham: CISRO Division of Wildlife and Ecology.

Faith, D. P. (2003). Environmental diversity (ED) as surrogate information for species-level diversity. *Ecography* **26**, 374–79.

Faith, D. P. & Walker, P. A. (1996a). Environmental diversity: on the best-possible use of surrogate data for assessing the relative biodiversity of sets of areas. *Biodiversity Conservation*, **5**, 399–415.

Faith, D. P. & Walker, P. A. (1996b). Integrating conservation and development: effective trade-offs between biodiversity and cost in the selection of protected areas. *Biodiversity Conservation*, **5**, 417–29.

Faith, D. P., Ferrier, S., & Walker, P. A. (2004). The ED strategy: how species-level surrogates indicate general biodiversity patterns through an "environmental diversity" perspective. *Journal of Biogeography*, **31**, 1207–17.

Faith, D. P., Margules, C. R. & Walker, P. A. (2001a). A biodiversity conservation plan for Papua New Guinea based on biodiversity trade-offs analysis. *Pacific Conservation Biology*, **6**, 304–24.

Faith, D. P., Margules, C. R., Walker, P. A., Stein, J. & Natera, G. (2001b). Practical applications of biodiversity surrogates and percentage targets for conservation in Papua New Guinea. *Pacific Conservation Biology*, **6**, 289–303.

Faith, D. P., Walker, P. A., Ive, J. R. & Belbin, L. (1996). Integrating conservation and forestry production: exploring trade-offs between biodiversity and production in regional land use assessment. *Forest Ecology and Management*, **85**, 251–60.

Faith, D. P., Walker, P. A. & Margules, C. R. (2001d). Some future prospects for systematic biodiversity planning in Papua New Guinea – and for biodiversity planning in general. *Pacific Conservation Biology*, **6**, 325–43.

Fandiño-Lozano, M. (1996). Framework for ecological evaluation oriented at the establishment and management of protected areas. (Ph.D. thesis, University of Amsterdam) ITC Publication No. 45, ITC, Enschede, The Netherlands, p. 195.

FAO (Food and Agriculture Organization of the United Nations). (1993). World soil resources: an explanatory note on the FAO world soil resources map at 1:25 000 000 scale. World soil resources report 66, rev. 1. FAO, Rome. Available from http://www.fao.org/sd/eidirect/gis/chap7.htm (accessed May 2003).

Ferrier, S. (2002). Mapping spatial pattern in biodiversity for regional conservation planning: where to from here? *Systematic Biology*, **51**, 331–63.

Ferrier, S., Drielsma, M., Manion, G. & Watson, G. (2002). Extended statistical approaches to modeling spatial pattern in biodiversity in north-east New South Wales. II. Community-level modelling. *Biodiversity and Conservation*, **11**, 2309–38.

Ferrier, S., Powell, G. V. N., Richardson, K. S. *et al.* (2004). Mapping more of terrestrial biodiversity for global conservation assessment. *BioScience*, **54**, 1101–9.

Ferrier, S., Pressey, R. L. & Barrett, T. W. (2000). A new predictor of the irreplaceability of areas for achieving a conservation goal, its application to real-world planning, and a research agenda for further refinement. *Biological Conservation*, **93**, 303–25.

Ferrier, S. & Watson, G. (1997). *An Evaluation of the Effectiveness of Environmental Surrogates and Modelling Techniques in Predicting the Distribution of Biological Diversity: Consultancy Report to the Biodiversity Convention and Strategy Section of the Biodiversity Group.* Canberra: Environment Australia.

Ferrier, S., Watson, G., Pearce, J. & Drielsma, M. (2002). Extended statistical approaches to modelling spatial pattern in biodiversity in north-east New South Wales. 1. Species-level modelling. *Biodiversity and Conservation*, **11**, 2275–307.

Fjeldsa, J. (1994). Geographical patterns for relict and young species of birds in Africa and South America and implications for conservation priorities. *Biodiversity and Conservation*, **3**, 207–26.

Flather, C. H., Wilson, K. R., Dean, D. J. & McComb, M. (1997). Identifying gaps in conservation networks: of indicators and uncertainty in geographic-based analyses. *Ecological Applications*, **7**, 532–42.

Fleishman, E., Blair, R. B. & Murphy, D. D. (2001). Empirical validation of a method for umbrella species selection. *Ecological Applications*, **11**, 1489–501.

Fleishman, E., Jonsson, B. G. & Sjoegren-Gulve, P. (2000). Focal species modeling for biodiversity conservation. *Ecological Bulletins*, **48**, 85–99.

Forman, R. T. T. (1995). *Land Mosaics: The Ecology of Landscapes and Regions.* Cambridge: Cambridge University Press.

Forsman, E. D., DeStefano, S., Raphael, M. G. & Gutiérrez, J. (1996). Demography of the northern spotted owl. *Studies in Avian Biology*, **17**, 1–122.

Funk, V. A., Richardson, K. S. & Ferrier, S. (2005). Survey-gap analysis in expeditionary research: where do we go from here? *Biological Journal of the Linnean Society*, **85**, 549–67.

Gadgil, M. & Guha, R. (1995). *Ecology and Equity: The Use and Abuse of Nature in Contemporary India.* New Delhi: Penguin.

Garshelis, D. L. (2000). Delusions in habitat evaluation: measuring use, selection, and importance. In *Research Techniques in Animal Ecology: Controversies and Consequences*, ed. L. Boitani & T. K. Fuller. New York: Columbia University Press, pp. 111–64.

Garson, J., Aggarwal, A. & Sarkar, S. (2002a). Birds as surrogates for biodiversity: an analysis of a data set from southern Québec. *Journal of Biosciences*, **27** (Suppl. 2), 347–60.

Garson, J., Aggarwal, A. & Sarkar, S. (2002b). *ResNet Ver 1.2 Manual.* University of Texas Biodiversity and Biocultural Conservation Laboratory, Austin, Texas. uts.cc.utexas.edu/~consbio/Cons/ResNet.html.

Gaston, K. J. (1994). *Rarity.* London: Chapman & Hall.

Gaston, K. J., Pressey, R. L. & Margules, C. R. (2002). Persistence and vulnerability: retaining biodiversity in the landscape and in protected areas. *Journal of Biosciences*, **27** (Suppl. 2), 361–84.

Geneletti, D. (2004). A GIS-based decision support system to identify nature conservation priorities in an alpine valley. *Land Use Policy*, **21**, 149–60.

Gibbons, D. W., Reid, J. B. & Chapman, R. A. (1993). *The New Atlas of Breeding Birds in Britain and Ireland: 1988–1991*. London: Poyser.

Giller, P. S. (1984). *Community Structure and Niche*. London: Chapman & Hall.

Gillison, A. N. & Brewer, K. R. W. (1985). The use of gradient directed transects or gradsects in natural resource survey. *Journal of Environmental Management*, **20**, 103–27.

Given, D. R. & Norton, D. A. (1993). A multivariate approach to assessing threat and for priority setting in threatened species conservation. *Biological Conservation*, **64**, 57–66.

Gleason, H. A. (1922). On the relation between species and area. *Ecology*, **3**, 158–62.

Goldblatt, P. & Manning, J. (2002). Plant diversity of the Cape Region of South Africa. *Annals of the Missouri Botanical Gardens*, **89**, 281–302.

Gole, C., Burton, M., Williams, K. J. *et al.* (2005). *Auction for Landscape Recovery*. Final Report. Commonwealth Market Based Instruments Program. WWF Australia, p. 184.

Goodman, D. (2002). Predictive Bayesian population viability analysis: a logic for listing criteria, delisting criteria, and recovery plans. In *Population Viability Analysis*, ed. S. R. Beissinger & D. R. McCullough. Chicago: University of Chicago Press, pp. 447–69.

Götmark, F. (1992). Naturalness as an evaluation criterion in nature conservation: a response to Anderson. *Conservation Biology*, **6**, 455–7.

Gower, J. C. (1966). Some distance properties in latent roots and vector methods used in multivariate analysis. *Biometrika*, **53**, 325–38.

Gower, J. (1971). Coefficients of association and similarity, based on binary (presence-absence) data: an evaluation. *Biometrics*, **27**, 857–71.

Graham, A., Margules, C., Beehler, B. & de Fretes, Y. (2005). *New Guinea Conservation Strategies*, Washington D. C.: Conservation International & CSIRO.

Groves, C. R., Jensen, D. B., Valutis, L. L. *et al.* (2002). Planning for biodiversity conservation: putting conservation science into practice. *BioScience*, **52**, 499–512.

Grubb, P. J., Kelly, D. & Mitchley, J. (1982). The control of relative abundance in communities of herbaceous plants. In *The Plant Community as a Working Mechanism*, ed. E. I. Newman. London: British Ecological Society Special Publication No. 1. Blackwell, pp. 79–97.

Guha, R. (1989). Radical American environmentalism and wilderness preservation: a third world critique. *Environmental Ethics*, **11**, 71–83.

Guisan, A., Edwards, T. C. & Hastie, T. (2002). Generalized linear and generalized additive models in studies of species distributions: setting the scene. *Ecological Modelling*, **157**, 89–100.

Haila, Y. & Margules, C. R. (1996). Survey research in conservation biology. *Ecography*, **19**, 323–31.

Halladay, P. & Gilmour, D. A. (eds.) (1995). *Conserving Biodiversity Outside Protected Areas: the Role of Traditional Agro-ecosystems*. Gland, Switzerland: IUCN.

Hampe, A. (2004). Bioclimate envelope models: what they detect and what they hide. *Global Ecology and Biogeography*, **13**, 469–70.

Hanski, I. A. (1998). Metapopulation dynamics. *Nature*, **396**, 41–9.

Hanski, I. A. (1999). *Metapopulation Ecology*. Oxford: Oxford University Press.

Harris, L. D. (1984). *The Fragmented Forest*. Chicago: The University of Chicago Press.

Harris, R. B., Metzgar, L. H. & Bevins, C. D. (1986). *GAPPS: generalized animal population projection system*. Version 3.0. Montana Cooperative Wildlife Research Unit, University of Montana, Missoula, Montana, USA.

Harvey, C. A., Tucker, N. I. J. & Estrada, A. (2004). Live fences, isolated trees, and windbreaks: tools for conserving biodiversity in fragmented tropical landscapes. In *Agroforestry and Biodiversity Conservation in Tropical Landscapes*, ed. G. Schroth, G. A. B. Da Fonseca, C. A. Harvey, *et al.* Washington: Island Press, pp. 261–89.

Hedrick, P. W., Lacy, R. C., Allendorf, F. W. & Soulé, M. E. (1996). Directions in conservation biology: comments on Caughley. *Conservation Biology*, **10**, 1312–20.

Herath, G. (2004). Incorporating community objectives in improved wetland management: the use of the analytic hierarchy process. *Journal of Environmental Management*, **70**, 263–73.

Higgs, A. J. (1981). Island biogeography and nature reserve design. *Journal of Biogeography*, **8**, 117–24.

Hilbert, D. W. & Muyzenberg, J. V. D. (1999). Using an artificial neural network to characterize the relative suitability of environments for forest types in a complex tropical vegetation mosaic. *Diversity and Distributions*, **5**, 263–74.

Hinloopen, E., Nijkamp, P. & Rietveld, P. (1983). The Regime method: a new multicriteria method. *Essays and Surveys on Multiple Criteria Decision Making*, ed. P. Hansen. Berlin: Springer, pp. 146–55.

Holdridge, L. R. (1967). *Life Zone Ecology*. San Jose: Tropical Science Center.

Holling, C. S. (ed.) (1978). *Adaptive Environmental Assessment and Management*. Chichester: Wiley.

Huang, W., Luukkanen, O., Johanson, S. *et al.* (2002). Agroforestry for biodiversity conservation of nature reserves: functional group identification and analysis. *Agroforestry Systems*, **55**, 65–72.

Hummel, M. (ed.) (1995). *Protecting Canada's Endangered Spaces*. Toronto: Key Porter Books.

Hurlbert, S. H. (1971). The non-concept of species diversity: a critique and alternative parameters. *Ecology*, **52**, 577–86.

Huston, M. A. (2005). The three phases of land use change: implications for biodiversity. *Ecological Applications*, **15**, 1864–78.

Hutchinson, G. E. (1958). Concluding remarks. Cold Spring Harbor Symposium. *Quantitative Biology*, **22**, 415–27.

Hutchinson, M. F. (1989). A new method for gridding elevation data with automatic removal of pits. *Journal of Hydrology*, **106**, 211–32.

Hutchinson, M. F. (1991). The application of thin plate smoothing splines to continent-wide data assimilation. In *Data Assimilation Systems*, ed. J. D. Jasper. Bureau of Meteorology Research Report No. 27, Bureau of Meteorology, Melbourne, pp. 104–13.

Hutchinson, M. F. (1993). Development of a continent-wide DEM with applications to terrain and climate analysis. In *Environmental Modelling with GIS*, ed. M. F. Goodchild, B. O. Parks & L. T. Steyaert. New York: Oxford University Press, pp. 392–9.

Hutchinson, M. F. (1995). Interpolating mean rainfall using thin plate smoothing splines. *International Journal of Geographical Information Systems*, **9**, 385–403.

Hutchinson, M. F. (1997). *ANUDEM Version 4.6*. Centre for Resource and Environmental Studies, Australian National University. cres.anu.edu.au/software/anudem.html.

Hutchinson, M. F., Belbin, L., Nicholls, A. O. *et al.* (1996). *BioRap. Volume 2. Spatial Modelling Tools*. Canberra: The Australian BioRap Consortium.

Huth, A., Drechsler, M. & Köhler, P. (2004). *Multicriteria evaluation of simulated logging scenarios in a tropical rain forest* Dordrecht: Kluwer.

Jannsen, R. (1992). *Multiobjective Decision Support for Environmental Management*. Dordrecht: Kluwer.

Janssen, R., Goosen, H., Verhoeven, M. *et al.* (2005). Decision support for integrated wetland management. *Environmental Modelling and Software*, **20**, 215–29.

Johnson, C. J. & Boyce, M. S. (2005). *A Quantitative Approach for Regional Environmental Assessment: Application of a Habitat-Based Population Viability Analysis to Wildlife of the Canadian Central Arctic*. Canadian Environmental Assessment Agency, Research & Development Monograph Series. http://www.ceaa-acee.gc.ca/015/0002/0028/index_e.htm

Justus, J. & Sarkar, S. (2002). The principle of complementarity in the design of reserve networks to conserve biodiversity: a preliminary history. *Journal of Biosciences*, **27** (Suppl. 2), 421–35.

Kamenetzky, R. D. (1982). The relationship between the analytic hierarchy process and the additive value function. *Decision Science*, **13**, 702–13.

Kangas, J. (1993). A multi-attribute preference model for evaluating the reforestation chain alternatives of a forest stand. *Forest Ecology and Management*, **59**, 271–88.

Kanowski, P. J., Gilmour, D. A., Margules, C. R. & Potter, C. S. (1999). *International Forest Conservation: Protected Areas and Beyond*. Canberra, ACT: Commonwealth of Australia.

Keeney, R. L. & Raiffa, H. (1993). *Decisions with Multiple Objectives: Preferences and Value Tradeoffs*. Cambridge: Cambridge University Press.

Keig, G. & Quigley, J. (1995). *Papua New Guinea Resource Information System. User's Guide. Version 2*. Canberra: Australian Agency for International Development.

Kerley, G. I. H., Pressey, R. L., Cowling, R. M., Boshoff, A. F. & Sims-Castley, R. (2003). Options for the conservation of large and medium-sized mammals in the Cape Floristic Region hotspot, South Africa. *Biological Conservation*, **112**, 169–90.

Kirkpatrick, J. B. (1983). An iterative method for establishing priorities for the selection of nature reserves: an example from Tasmania. *Biological Conservation*, **25**, 127–34.

Kirkpatrick, J. B. & Brown, M. J. (1994). A comparison of direct and environmental domain approaches to planning reservation of forest higher plant communities and species in Tasmania. *Conservation Biology*, **8**, 217–24.

Kirkpatrick, J. B. & Harwood, C. E. (1983). Conservation of Tasmanian macrophytic wetland vegetation. *Proceedings of the Royal Society of Tasmania*, **117**, 5–20.

Kirkpatrick, S., Gelatt, C. D. & Vecchi, M. P. (1983). Optimisation by simulated annealing. *Science*, **220**, 671–80.

Knight, A. T., Driver, A., Cowling, R. M. *et al.* (2006). Designing systematic conservation assessments that promote effective implementation: best practice from South Africa. *Conservation Biology*, **20**, 739–50.

Krebs, C. J. (2001). *Ecology. The Experimental Analysis of Distribution and Abundance*, 5th edn. San Francisco: Addison Wesley Longman.

Kuusipalo, J. & Kangas, J. (1994). Managing biodiversity in a forestry environment. *Conservation Biology*, **8**, 450–60.

Lacy, R. C. (1997). Importance of genetic variation to the viability of mammalian populations. *Journal of Mammalogy*, **78**, 320–35.

Lambeck, R. J. (1997). Focal species: a multi-umbrella for nature conservation. *Conservation Biology*, **11**, 849–56.

Lambeck, R. J. (1999). *Landscape Planning for Biodiversity Conservation in Agricultural Regions: A Case Study from the Wheatbelt of Western Australia. Biodiversity Technical Paper 2*, Canberra: Environment Australia.

Lande, R., Engen, S. & Sæther, B. E. (2003). *Stochastic Population Dynamics in Ecology and Conservation*. Oxford: Oxford University Press.

Landres, P. B., Verner, J. & Thomas, J. W. (1988). Ecological uses of vertebrate indicator species: a critique. *Conservation Biology*, **2**, 316–28.

Laukkanen, S., Kangas, A. & Kangas, J. (2002). Applying voting theory in natural resource management: a case of multiple-criteria group decision support. *Journal of Environmental Management*, **64**, 127–37.

Laut, P., Heyligers, P. C., Keig, G. *et al.* (1977). *Environments of South Australia: Handbook*. Canberra: CSIRO Division of Land Use Research.

Lawler, J. J. & Schumaker, N. H. (2004). Evaluating habitat as a surrogate for population viability using a spatially explicit population model. *Environmental Monitoring and Assessment*, **94**, 85–100.

Lawton, J. H., Prendergast, J. R. & Eversham, B. C. (1994). The numbers and spatial distributions of species: analyses of British data. In *Systematics and Conservation Evaluation*, ed. P. L. Forey, C. J. Humphries & R. I. Vane-Wright. Oxford: Oxford University Press, pp. 177–95.

Leader-Williams, N., Harrison, J. & Green, M. J. B. (1990). Designing protected areas to conserve natural resources. *Science Progress*, **74**, 189–204.

Leathwick, J. R. & Mitchell, N. D. (1992). Forest pattern, climate and vulcanism in central North Island, New Zealand. *Journal of Vegetation Science*, **3**, 603–16.

Lee, J. T. & Thompson, S. (2005). Targeting sites for habitat creation: an investigation into alternative scenarios. *Landscape and Urban Planning*, **71**, 17–28.

Lewis, M. (2003). *Inventing Global Ecology: Tracking the Biodiversity Ideal in India, 1945–1997*. New Delhi: Orient Longman.

Li, W., Wang, Z. & Tang, H. (1999). Designing the buffer zone of a nature reserve: a case study in Yancheng Biosphere Reserve, China. *Biological Conservation*, **90**, 159–65.

Lindenmayer, D. & Burgman, M. (2005). *Practical Conservation Biology*. Melbourne: CSIRO.

Lindenmayer, D. B., Cunningham, R. B., Tanton, M. T. & Smith, A. P. (1990). The conservation of arboreal marsupials in montane ash forests of the Central Highlands of Victoria, south-east Australia. I. The loss of trees with hollows and its implications for the conservation of Leadbeater's possum, *Gymnobelideus leadbeateri* McCoy (Marsupialia: Petauridae). *Biological Conservation*, **54**, 133–45.

Lindenmayer, D. B., Manning, A. D., Smith, P. L. *et al.* (2002). The focal-species approach and landscape restoration: a critique. *Conservation Biology*, **16**, 338–45.

Lindenmayer, D. B. & Possingham, H. P. (1996). Ranking conservation and timber management options for Leadbeater's possum in south-eastern Australia using population viability analysis. *Conservation Biology*, **10**, 235–51.

Lindo systems. (1995). LINDO/*386, 5.3*. Chicago: Lindo Systems, Inc.

Lombard, A. T., Cowling, R. M., Pressey, R. L. & Rebelo, A. G. (2003). Effectiveness of land classes as surrogates for species in conservation planning for the Cape Floristic Region. *Biological Conservation*, **112**, 45–62.

Lowry, D. C. & Jennings, J. N. (1974). The Nullarbor Karst, Australia. *Zeit. Geomorph. N. F*, **18**, 36–81.

Lund, M. & Rahbek, C. (2002). Cross-taxon congruence in complementarity and conservation of temperate biodiversity. *Animal Conservation*, **5**, 163–71.

MacArthur, R. H. & Wilson, E. O. (1967). *The Theory of Island Biogeography*. New Jersey: Princeton University Press.

Mackey, B. G., Nix, H. A., Stein, J. A., Cork, S. E. & Bullen, F. T. (1989). Assessing the representativeness of the wet tropics of Queensland World Heritage property. *Biological Conservation*, **50**, 279–303.

Malczewski, J., Moreno-Sánchez, R., Bojórquez-Tapia, L. & Ongay-Delhumeau, E. (1997). Multicriteria group decision-making model for environmental conflict analysis in the Cape Region, Mexico. *Journal of Environmental Planning and Management*, **40**, 349–74.

Margules, C. R. (1989). Introduction to some Australian developments in conservation evaluation. *Biological Conservation*, **50**, 1–11.

Margules, C. R. & Austin, M. P. (1994). Biological models for monitoring species decline: the construction and use of data bases. *Philosophical Transactions of the Royal Society of London Series B*, **343**, 69–75.

Margules, C. R., Higgs, A. J. & Rafe, R. W. (1982). Modern biogeographic theory: are there lessons for nature reserve design? *Biological Conservation*, **24**, 115–28.

Margules, C. R., Nicholls, A. O. & Austin, M. P. (1987). Diversity of *Eucalyptus* species predicted by a multivariable environmental gradient. *Oecologia*, **71**, 229–32.

Margules, C. R., Nicholls, A. & Pressey, R. L. (1988). Selecting networks of reserves to maximise biological diversity. *Biological Conservation*, **43**, 63–76.

Margules, C. R. & Nicholls, A. O. (1987). Assessing the conservation value of remnant habitat "islands": mallee patches on the western Eyre Peninsula, South Australia. In *Nature Conservation: the Role of Remnants of Native Vegetation*, ed. D. A. Saunders, G. W. Arnold, A. A. Burbidge & A. J. M. Hopkins. Sydney: Surrey Beatty & Sons, pp. 89–102.

Margules, C. R. & Pressey, R. L. (2000). Systematic conservation planning. *Nature*, **405**, 243–53.

Margules, C. R., Pressey, R. L. & Nicholls, A. O. (1991). Selecting nature reserves. In *Nature Conservation: Cost Effective Biological Surveys and Data Analysis*, ed. C. R. Margules & M. P. Austin. Melbourne: CSIRO, 90–7.

Margules, C. R., Pressey, R. L. & Williams, P. H. (2002). Representing biodiversity: data and procedures for identifying priority areas for conservation. *Journal of Biosciences*, **27** (Suppl. 2), 309–26.

Margules, C. R., Redhead, T. D., Faith, D. P. & Hutchinson, M. F. (1995). *BioRap: Guidelines for using the BioRap Methodology and Tools*. Canberra: CSIRO.

Margules, C. R. & Stein, J. L. (1989). Patterns in the distributions of tree species and the selection of nature reserves: an example from *Eucalyptus* forests in south-eastern New South Wales. *Biological Conservation*, **50**, 219–38.

Margules, C. R. & Usher, M. B. (1981). Criteria used in assessing wildlife conservation potential: a review. *Biological Conservation*, **21**, 79–109.

Mayr, E. (1957). Species concepts and definitions. *American Association for the Advancement of Science Publications*, **50**, 1–22.

McAlpine, J. R., Keig, G. & Falls, R. (1983). *Climate of Papua New Guinea*. Canberra: Australian National University Press.

McAlpine, J. R., Keig, G. & Short, K. (1975). *Climatic Tables for Papua New Guinea. Division of Land Use Research Technical Paper No. 37*. Melbourne: CSIRO.

McCarthy, M. A., Andelman, S. J. & Possingham, H. P. (2003). Reliability of relative predictions in population viability analysis. *Conservation Biology*, **17**, 982–9.

McIntyre, S. (1992). Risks associated with the setting of conservation priorities from rare plant species lists. *Biological Conservation*, **60**, 31–7.

McKenzie, N. L. & Robinson, A. C. (eds.) (1987). *A Biological Survey of the Nullarbor Region South and Western Australia in 1984*. Adelaide: South Australian Department of Environment and Planning.

McKenzie, N. L., Belbin, L., Margules, C. R. & Keighery, G. J. (1989). Selecting representative reserve systems in remote areas: a case study in the Nullarbor region, Australia. *Biological Conservation*, **50**, 239–61.

McKenzie, N. L., Johnston, R. B. & Kendrick, P. G. (eds.) (1991). *Kimberley Rainforests of Australia*. Chipping Norton, NSW: Surrey Beatty & Sons.

McKinney, M. L. (2002). Urbanization, biodiversity and conservation. *BioScience*, **52**, 883–90.

McLachlan, A. & Burns, M. (1992). Headland bypass dunes on the South African coast: 100 years of (mis)management. In *Coastal Dunes: Geomorphology, Ecology and Management for Conservation*, ed. R. W. G. Carter, T. G. F. Curtis & M. J. Sheehy-Skeffington. Rotterdam: A. A. Balkema, pp. 71–9.

McNeely, J. A. & Scherr, S. J. (2003). *Ecoagriculture: Strategies to Feed the World and Save Wild Biodiversity*. Washington D.C.: Island Press, p. 323.

Meffe, G. K. & Carroll, C. R. (1994). *Principles of Conservation Biology*. Sunderland, MA: Sinauer Associates.

Mendoza, G. & Prabhu, R. (2000). Multiple criteria decision making approaches to assessing forest sustainability using criteria and indicators: a case study. *Forest Ecology and Management*, **131**, 107–26.

Mendoza, G. & Sprouse, W. (1989). Forest planning and decision making under fuzzy environments: an overview and illustration. *Forest Science*, **35**, 481–502.

Millett, I. & Wedley, W. (2002). Modelling risk and uncertainty in the Analytic Hierarchy Process. *Journal of Multi-Criteria Decision Analysis*, **11**, 97–107.

Millsap, B. A., Gore, J. A., Runde, D. E. & Cerulean, S. I. (1990). Setting priorities for the conservation of fish and wildlife species in Florida. *Wildlife Monographs*, **111**, 1–57.

Moffett, A., Dyer, J. S., & Sarkar, S. (2006). Integrating biodiversity representation with multiple criteria in North-Central Namibia using non-dominated alternatives and a modified analytic hierarchy process. *Biological Conservation*, **129**, 181–91.

Moffett, A. & Sarkar, S. (2005). Incorporating multiple criteria into the design of conservation area networks: a minireview with recommendations. *Diversity and Distributions*, **12**, 125–37.

Moilanen, A. & Cabeza, M. (2002). Single-species dynamic site selection. *Ecological Applications*, **12**, 913–26.

Molloy, J. & Davis, A. (1992). *Setting Priorities for the Conservation of New Zealand's Threatened Plants and Animals*. Wellington: Department of Conservation.

Moritz, C., Richardson, K. S., Ferrier, S. *et al.* (2001). Biogeographical concordance and efficiency of taxon indicators for establishing conservation priority in a tropical rainforest biota. *Proceedings of the Royal Society London Series B*, **268**, 1875–81.

Morris, W. F. & Doak, D. F. (2002). *Quantitative Conservation Biology: An introduction to Population Viability Analysis*. Sunderland: Sinauer.

Mueller-Dombois, D. & Ellenberg, H. (1974). *Aims and Methods of Vegetation Ecology*. New York, USA: John Wiley and Sons, p. 547.

Munton, P. (1987). Concepts of threat to the survival of species used in Red Data books and similar compilations. In *The Road to Extinction: Problems with Categorizing the Status of Taxa Threatened with Extinction*, ed. R. Fitter & M. Fitter. Gland: IUCN, pp. 71–111.

Myers, N. (1988). Threatened biotas: "hot-spots" in tropical forests. *The Environmentalist*, **8**, 187–208.

Myers, N. (1990). The biodiversity challenge: expanded hot-spots analysis. *The Environmentalist*, **10**, 243–56.

Myers, N., Mittermeier, R. A., Mittermeier, C. G., da Fonseca, G. A. & Kent, J. (2000). Biodiversity hotspots for conservation priorities. *Nature*, **403**, 853–8.

New, M., Hulme, M. & Jones, P. (1999). Representing twentieth-century space-time climate variability. Part I: Development of a 1961–90 mean monthly terrestrial climatology. *Journal of Climate*, **12**, 829–56.

Nicholls, A. O. (1989). How to make biological surveys go further with generalised linear models. *Biological Conservation*, **50**, 51–75.

Nicholls, A. O. (1991). Examples of the use of generalised linear models in analysis of survey data for conservation evaluation. In *Nature Conservation: Cost Effective Biological Surveys and Data Analysis*, ed. C. R. Margules & M. P. Austin. Melbourne: CSIRO, pp. 54–63.

Nicholls, A. O. (1998). Integrating population abundance, dynamics and distribution into broad scale priority setting. In *Conservation in a Changing World: Integrating Processes into Priorities for Action*, ed. G. M. Mace, A. Balmford & J. Ginsberg. Cambridge: Cambridge University Press.

Nicholls, A. O. & Margules, C. R. (1993). An upgraded reserve selection algorithm. *Biological Conservation*, **64**, 165–9.

Nix, H. A. (1986). A biogeographic analysis of Australian elapid snakes. In *Atlas of Elapid Snakes of Australia*. *Australian Flora and Fauna Series No. 11*, ed. R. Longmore. Canberra: Australian Government Publishing Service, pp. 4–17.

Nix, H. A. (2004). Inverting the paradigm. *Pacific Conservation Biology*, **10**, 76.

Nix, H. A., Faith, D. P., Hutchinson, M. F. *et al.* (2000). *The BioRap Toolbox. A National Study of Biodiversity Assessment and Planning for Papua New Guinea*, Canberra: Centre for Resource and Environment Studies, Australian National University.

Nix, H. A. & Switzer, M. A. (1991). Rainforest Animals: Atlas of Vertebrates Endemic to Australia's Wet Tropics. *Kowari* **1**. (Canberra: Australian National Parks and Wildlife Service.)

Norton, B. G. (1987). *Why Preserve Natural Variety?* Princeton: Princeton University Press.

Norton, B. G. (2003). *Searching for Sustainability*. New York: Cambridge University Press.

Noss, R., Carroll, C., Vance-Borland, K. & Wuerthner, G. (2002). A multicriteria assessment of the irreplaceability and vulnerability of sites in the Greater Yellowstone Ecosystem. *Conservation Biology*, **16**, 895–908.

Noss, R. F., Strittholt, J. R., Vance-Borland, K., Carroll, C. & Frost, P. (1999). A conservation plan for the Klamath-Siskiyou ecoregion. *Natural Areas Journal*, **19**, 392–411.

Novikova, N. & Pospelova, I. (2002). Multicriteria decision making under uncertainty. *Mathematical Programming*, **92**, 537–54.

Ødegaard, F. (2000). How many species of arthropods? Erwin's estimate revised. *Biological Journal of the Linnean Society*, **71**, 583–97.

Ohsawa, M. (1987). *Life Zone Ecology of Bhutan Himalaya*. Chiba University, Chiba: Laboratory of Ecology.

Oliver, I., Beattie, A. J. & York, A. (1998). Spatial fidelity of plant, vertebrate, and invertebrate assemblages in multiple-use forest in eastern Australia. *Conservation Biology*, **12**, 822–35.

Oliver, I. & Parkes, D. (2003). *A Prototype Toolkit for Predicting the Biodiversity Benefits (and Disbenefits) of Land Use Change, Version 4.0*. Centre for Natural Resources, New South Wales Department of Land and Water Conservation, Parramatta Sydney: NSW Government. Unpublished.

Olson, D. M., Dinerstein, E., Wikramanayake, E. D. *et al.* (2001). Terrestrial ecoregions of the world: a new map of life on Earth. *BioScience*, **51**, 933–8.

O' Riordan, T., Fairbrass, J., Welp, M. & Stoll-Kleeman, S. (2002). The politics of biodiversity in Europe. *Biodiversity, Sustainability and Human Communities*, **71**, 115–41.

Palmer, M. W. (1990). The estimation of species richness by extrapolation. *Ecology*, **71**, 1195–8.

Panetta, E. D. & Dodd, J. (1987). Factors determining seed persistence of *Chondrilla juncea* L. (skeleton weed) in southern Western Australia. *Australian Ecology*, **13**, 211–24.

Panzer, R., & Schwartz, M. W. (1998). Effectiveness of a vegetation based approach to invertebrate conservation. *Conservation Biology*, **12**, 693–702.

Parkes, D., Newell, G. & Cheal, D. (2003). Assessing the quality of native vegetation: The 'habitat hectares' approach. *Ecological Management and Restoration*, **4** (Suppl.), S29–S38.

Pausas, J. G., Austin, M. P. & Noble, I. R. (1997). A forest simulation model for predicting Eucalypt dynamics and habitat quality for arboreal marsupials. *Ecological Applications*, **7**, 921–33.

Pearson, R. G. & Dawson, T. P. (2003). Predicting the impacts of climate change on the distribution of species: are climate envelope models useful? *Global Ecology and Biogeography*, **12**, 361–71.

Pereira, J. & Duckstein, L. (1993). A multiple criteria decision-making approach to GIS-based land suitability evaluation. *International Journal of Geographical Information Systems*, **7**, 407–24.

Perring, F. (1958). A theoretical approach to a study of chalk grasslands. *Journal of Ecology*, **46**, 665–79.

Perring, F. (1959). Topographical gradients in chalk grassland. *Journal of Ecology*, **47**, 447–81.

Perrow, M. R. & Davy, A. J. (eds.) (2002). *Handbook of Ecological Restoration*. Cambridge: Cambridge University Press.

Peterson, A. T., Soberón, J. & Sánchez-Cordero, V. (1999). Conservatism of ecological niches in evolutionary time. *Science*, **285**, 1265–7.

Pharo, E. J., Beattie, A. J. & Pressey, R. I. (2000). Effectiveness of using vascular plants to select reserves for bryophytes and lichens. *Biological Conservation*, **96**, 371–8.

Phillips, S. J., Anderson, R. P. & Schapire, R. E. (2006). Maximum entropy modeling of species geographic distributions. *Ecological Modelling*, **190**, 231–59.

Phillips, S. J., Dudik, M. & Shapire, R. E. (2004). A maximum entropy approach to species distribution modeling. In *Proceedings of the 21st International Conference on Machine Learning*. New York: ACM Press, pp. 655–62.

Phua, M.-H. & Minowa, M. (2005). A GIS-based multi-criteria decision making approach to forest conservation planning at a landscape scale: a case study of the Kinabalu area, Sabah, Malaysia. *Landscape and Urban Planning*, **71**, 207–22.

Picker, M. D. & Samways, M. J. (1996). Faunal diversity and endemicity of the Cape Peninsula, South Africa – a first assessment. *Biodiversity and Conservation*, **5**, 591–606.

Pickett, S. T. A. & Thompson, J. A. (1978). Patch dynamics and the design of nature reserves. *Biological Conservation*, **13**, 27–37.

Pierce, S. M., Cowling, R. M., Knight A. T. *et al.* (2005). Systematic conservation planning products for land-use planning: interpretation for implementation. *Biological Conservation*, **125**, 441–58.

Podger, F. D., Mummery, D. C., Palzer, C. R. & Brown, M. J. (1990). Bioclimatic analysis of the distribution of damage to native plants in Tasmania by *Phytophthora cinnamomi*. *Australian Journal of Ecology*, **15**, 281–90.

Poon, E. L. & Margules, C. R. (2004). Searching for new populations of rare plant species in remote locations. In *Sampling Rare or Elusive Species. Concepts, Designs and Techniques for Estimating Population Parameters*, ed. W. L. Thompson. Washington: Island Press, pp. 189–210.

Possingham, H., Ball, I. & Andelman, S. (2000). Mathematical methods for identifying representative reserve networks. In *Quantitative Methods for Conservation Biology*, ed. S. Ferson & M. Burgman. New York: Springer-Verlag, pp. 291–305.

Possingham, H. P. & Davies, I. (1995). ALEX: a model for the viability analysis of spatially structured populations. *Biological Conservation*, **73**, 143–50.

Possingham, H. P., Lindenmayer, D. B. & Norton, T. W. (1993). A framework for the improved management of threatened species based on population viability analysis. *Pacific Conservation Biology*, **1**, 39–45.

Possingham, H. P., Wilson, K. A., Andelman, S. J. & Vynne, C. H. (2006). Protected areas: goals, limitations, and design. In *Principles of Conservation Biology*, ed. M. J. Groom, G. K. Meffe & C. R. Carroll, 3rd edn. Sunderland, MA: Sinauer Associates Inc., pp. 509–33.

Powell, G. N. & Bjork, R. (1995). Implications of intratropical migration on reserve design: a case study using *Pharomachrus mocinno*. *Conservation Biology*, **9**, 354–62.

Power, M. E., Tilman, D., Estes, J. A. *et al.* (1996). Challenges in the quest for keystones. *BioScience*, **46**, 609–20.

Prendergast, H. D. V. & Hattersley, P. W. (1985). Distribution and cytology of Australian *Neurachne* and its allies (Poaceae), a group containing C3, C4 and C3–C4 intermediate species. *Australian Journal of Botany*, **33**, 317–36.

Prendergast, J. R., Quinn, R. M. & Lawton, J. H. (1999). The gaps between theory and practice in selecting nature reserves. *Conservation Biology*, **13**, 484–92.

Prendergast, J. R., Quinn, R. M., Lawton, J. H., Eversham, B. C. & Gibbons, D. W. (1993). Rare species, the coincidence of diversity hotspots and conservation strategies. *Nature (London)*, **365**, 335–7.

Pressey R. L. (1990) Reserve selection in New South Wales: where to from here? *Australian Zoologist*, **26**, 70–5.

Pressey, R. L. (1992). Nature conservation in rangelands: lessons from research on reserve selection in New South Wales. *Rangeland Journal*, **14**, 214–26.

Pressey, R. L. (1994). *Ad Hoc* Reservations: forward or backward steps in developing representative reserve systems. *Conservation Biology*, **8**(3), 662–8.

Pressey, R. L. (1998). Algorithms, politics and timber: an example of the role of science in a public political negotiation process over new conservation areas in production forests. In *Ecology for Everyone: Communicating Ecology to Scientists, the Public and Politicians*, ed. R. Willis & R. Hobbs. Sydney: Surrey Beatty, pp. 73–87.

Pressey, R. L. (1999). Applications of irreplaceability analysis to planning and management problems. *Parks*, **9**, 42–51.

Pressey, R. L. & Cowling, R. M. (2001). Reserve selection algorithms and the real world. *Conservation Biology*, **15**, 275–7.

Pressey, R. L., Cowling, R. M. & Rouget, M. (2003). Formulating conservation targets for biodiversity pattern and process in the Cape Floristic Region, South Africa. *Biological Conservation*, **112**, 99–127.

Pressey, R. L., Ferrier, S., Hager, T. C. *et al.* (1996). How well protected are the forests of north-eastern New South Wales? – analyses of forest environments in relation to tenure, formal protection measures and vulnerability to clearing. *Forest Ecology and Management*, **85**, 311–33.

Pressey, R. L., Humphries, C. J., Margules, C. R., Vane-Wright, R. I. & Williams, P. H. (1993). Beyond opportunism: key principles for systematic reserve selection. *Trends in Ecology and Evolution*, **8**(4), 124–8.

Pressey, R. L., Johnson, I. R. & Wilson, P. D. (1994). Shades of irreplaceability: towards a measure of the contribution of sites to a reservation goal. *Biodiversity and Conservation*, **3**, 242–62.

Pressey, R. L. & Nicholls, A. O. (1989). Efficiency in conservation evaluation: scoring versus iterative approaches. *Biological Conservation*, **50**, 199–218.

Pressey, R. L., Possingham, H. P. & Day, J. R. (1997). Effectiveness of alternative heuristic algorithms for approximating minimum requirements for conservation reserves. *Biological Conservation*, **80**, 207–19.

Pressey, R. L., Possingham, H. P. & Margules, C. R. (1996). Optimality in reserve selection algorithms: when does it matter and how much? *Biological Conservation*, **76**, 259–67.

Pressey, R. L. & Taffs, K. H. (2001). Scheduling conservation action in production landscapes: priority areas in western New South Wales defined by irreplaceability and vulnerability to vegetation loss. *Biological Conservation*, **100**, 355–76.

Pulliam, H. R. (1988). Sources, sinks, and population regulation. *American Naturalist*, **132**, 652–61.

Rabinowitz, D., Cairns, S. & Dillon, T. (1986). Seven forms of rarity and their frequency in the flora of the British Isles. In *Conservation Biology: The Science of Scarcity and Diversity*, ed. M. E. Soulé. Sunderland: Sinauer, pp. 182–204.

Rebelo, A. G. & Siegfried, W. R. (1990). Protection of fynbos vegetation: ideal and real-world options. *Biological Conservation*, **54**, 15–31.

Redford, K., Andrews, M., Brown, D. *et al.* (1997). *Designing a Geography of Hope: Guidelines for Ecoregion Conservation in The Nature Conservancy*. Arlington: The Nature Conservancy.

Redpath, S., Arroyo, B., Leckie, F. *et al.* (2004). Using decision modeling with stakeholders to reduce human–wildlife conflict: a raptor-grouse case study. *Conservation Biology*, **18**, 350–9.

Reed, J. M. (1992). A system for ranking conservation priorities for neotropical migrant birds based on relative susceptibility to extinction. In *Ecology and Conservation of Neotropical Migrant Landbirds*, ed. J. M. Hagan III & D. W. Johnston. Washington, D.C.: Smithsonian Institution Press, pp. 524–36.

Reed, J. M., Mills, L. S., Dunning Jr., J. B. *et al.* (2002). Emerging issues in population viability analysis. *Conservation Biology*, **16**, 7–19.

Regan, H., Colyvan, M. and Burgman, M. A. (2002). A taxonomy and treatment of uncertainty for ecology and conservation biology. *Ecological Applications*, **12**, 618–28.

Reyers, B., van Jaarsveld, A. S. & Kruger, M. (2000). Complementarity as a biodiversity indicator strategy. *Proceedings of the Royal Society London B*, **267**, 505–13.

Richards, B. N., Bridges, R. G., Curtin, R. A. *et al.* (1990). *Biological Conservation of the South-Eastern Forests. (Report of the Joint Scientific Committee)*. Canberra: Australian Government Publishing Service.

Richardson, B. J. & McKenzie, A. M. (1992). Australia's biological collections and those who use them. *Australian Biologist*, **5**, 1930.

Richardson, K. S. & Funk V. A. (1999). An approach to designing a systematic protected area system in Guyana. *Parks*, **9**, 7–16.

Ride, W. D. L. (1975). Towards an integrated system: a study of selection and acquisition of National Parks and nature reserves in Western Australia. In *A National System of Ecological Reserves in Australia*, ed. F. Fenner. Canberra: Australian Academy of Science, pp. 64–85.

Robert, C. P. (2001). *The Bayesian Choice*. Berlin: Springer.

Robinson, A. C., McKenzie, N. L. & Davey, A. G. (1987). Conclusion and conservation recommendations. In *A Biological Survey of the Nullarbor Region, South and Western Australia in 1984*, ed. N. L. McKenzie & A. C. Robinson. Adelaide: South Australian Department of Environment & Planning, pp. 233–41.

Rodrigues, A. S., Cerdeira, J. O. & Gaston, K. J. (2000). Flexibility, efficiency, and accountability: adapting reserve selection algorithms to more complex conservation problems. *Ecography*, **23**, 565–74.

Rodrigues, A. S. & Gaston, K. J. (2002). Optimisation in reserve selection procedures – why not? *Biological Conservation*, **107**, 123–9.

Rosenzweig, M. L. (2003). *Win-Win Ecology: How the Earth's Species can Survive in the Midst of Human Enterprise*. Oxford: Oxford University Press.

Rothley, K. D. (1999). Designing bioreserve networks to satisfy multiple, conflicting demands. *Ecological Applications*, **9**, 741–50.

Rouget, M., Cowling, R. M., Lombard, A. T., Knight, A. T. & Kerley, G. I. H. (2006). Designing large-scale corridors for pattern and process. *Conservation Biology*, **20**, 549–61.

Rouget, M., Cowling, R. M., Pressey, R. L. & Richardson, D. M. (2003a). Identifying spatial components of ecological and evolutionary processes for regional conservation planning in the Cape Floristic Region, South Africa. *Diversity and Distributions*, **9**, 191–210.

Rouget, M., Richardson, D. M. & Cowling, R. M. (2003b). The current configuration of protected areas in the Cape Floristic Region, South Africa – reservation bias and representation of biodiversity patterns and processes. *Biological Conservation*, **112**, 129–45.

Rouget, M., Richardson, D. M., Cowling, R. M., Lloyd, W. J. & Lombard, A. T. (2003c). Current patterns of habitat transformation and future threats to biodiversity in terrestrial ecosystems of the Cape Floristic Region, South Africa. *Biological Conservation*, **112**, 63–85.

Ryti, R. (1992). Effect of the focal taxon on the selection of nature reserves. *Ecological Applications*, **2**, 404–10.

Saaty, T. (1980). *The Analytic Hierarchy Process: Planning, Priority Setting, Resource Allocation*. New York: McGraw-Hill Inc.

Sarakinos, H., Nicholls, A. O., Tubert, A. *et al.* (2001). Area prioritization for biodiversity conservation in Québec on the basis of species distributions: a preliminary analysis. *Biodiversity Conservation*, **10**, 1419–72.

Sarkar, S. (1998). *Genetics on Reductionism*. New York: Cambridge University Press.

Sarkar, S. (1999). Wilderness preservation and biodiversity conservation – keeping divergent goals distinct. *BioScience*, **49**, 405–12.

Sarkar, S. (2002). Defining "biodiversity": assessing biodiversity. *Monist*, **85**, 131–55.

Sarkar, S. (2003). Conservation area networks. *Conservation and Society*, **1**, v–vii.

Sarkar, S. (2004). Conservation Biology. In *The Stanford Encyclopedia of Philosophy*, summer 2004 edn., ed. E. N. Zalta. plato.stanford.edu/archives/sum2004/entries/conservation-biology/.

Sarkar, S. (2005). *Biodiversity and Environmental Philosophy: An Introduction*. New York: Cambridge University Press.

Sarkar, S., Aggarwal, A., Garson, J., Margules, C. R. & Zeidler, J. (2002). Place prioritization for biodiversity content. *Journal of Biosciences*, **27** (Suppl. 2), 339–46.

Sarkar, S. & Garson, J. (2004). Multiple criterion synchronization for conservation area network design: the use of non-dominated alternative sets. *Conservation and Society*, **2**, 433–48.

Sarkar, S., Justus, J., Fuller, T. *et al.* (2005). Effectiveness of environmental surrogates for the selection of conservation area networks. *Conservation Biology* **19**, 815–25.

Sarkar, S. & Margules, C. R. (2002). Operationalizing biodiversity for conservation planning. *Journal of Biosciences*, **27** (Suppl. 2), 299–308.

Sarkar, S., Moffett, A., Sierra, R. *et al.* (2004a). Incorporating multiple criteria into the design of conservation area networks. *Endangered Species Update*, **21**, 100–7.

Sarkar, S., Pappas, C., Garson, J., Aggarwal, A. & Cameron, S. (2004b). Place prioritization for biodiversity conservation using probabilistic surrogate distribution data. *Diversity and Distributions*, **10**, 125–33.

Sarkar, S., Parker, N. C., Garson, J., Aggarwal, A. & Haskell, S. (2000). Place prioritization for Texas using GAP data: the use of biodiversity and environmental surrogates within socio-economic constraints. *GAP Analysis Program Bulletin*, **9**, 48–50.

Sarkar, S., Pressey, R. L., Faith, D. P. *et al.* (2006). Biodiversity conservation planning tools: present status and challenges for the future. *Annual Review of Environment and Resources*, in press.

Sattler, P. S. & Williams, R. D. (eds.) (1999). *The Conservation Status of Queenslands Bioregional Ecosystems*. Brisbane, Australia: Environmental Protection Agency.

Sauer, J. D. (1969). Oceanic islands and biogeographical theory. *Geographical Review*, **59**, 582–93.

Saunders, D. A., Hobbs, R. J. & Margules, C. R. (1991). Biological consequences of ecosystem fragmentation: a review. *Conservation Biology*, **5**, 18–32.

Schmoldt, D. L., Kangas, J., Mendoza, G. A. & Pesonen, M. (eds.) (2001). *The Analytic Hierarchy Process in Natural Resource and Environmental Decision Making*. Managing Forest Ecosystems Vol. 3. Drodrecht: Kluwer Academic Publishers.

Schultz, S. M., Dunham, A. E., Root, K. V. *et al.* (1999). *Conservation Biology with RAMAS® EcoLab*. Sunderland, MA: Sinauer Associates.

Schumaker, N. H. (1998). A user's guide to the PATCH model. EPA/600/R-98/135. US Environmental Protection Agency, Environmental Research Laboratory. www.epa.gov/naaujydh/pages/models/patch/patchmain.htm.

Scott, J. M., Heglund, P. J., Hauffer, J. B. *et al.* (2002). *Predicting Species Occurrences: Issues of Accuracy and Scale*. Covelo, California: Island Press.

Shaffer, M. L. (1978). Determining minimum viable population sizes: a case study of the grizzly bear. Unpublished Ph.D. thesis, Duke University, Durham.

Shrader-Frechette, K. S. & McCoy, E. D. (1993). *Method in Ecology: Strategies for Conservation*. Cambridge: Cambridge University Press.

Sierra, R. (ed.) (1999). *Propuesta Preliminar de un Sistema de Clasificacion de Vegetacion para el Ecuador Continental*. Quito: Proyecto INEFAN/GEF-BIRF y EcoCienca.

Sierra, R., Campos, F. & Chamberlin, J. (2002). Assessing biodiversity conservation priorities: ecosystem risk and representativeness in continental Ecuador. *Landscape and Urban Planning*, **59**, 95–110.

Simberloff, D. A. (1988). The contribution of population and community biology to conservation science. *Annual Reviews of Ecology and Systematics*, **19**, 473–511.

Smith, P. G. R. & Theberge, J. B. (1986). A review of criteria for evaluating natural areas. *Environmental Management*, **10**, 715–34.

Sneath, P. H. A. & Sokal, R. R. (1973). *Numerical Taxonomy. The Principles and Practice of Numerical Classification.* San Francisco: W. H. Freeman.

Soulé, M. E. (1990). The real work of systematics. *Annals of the Missouri Botanical Garden*, **77**, 4–12.

Soulé, M. E. & Sanjayan, M. A. (1998). Conservation targets: do they help? *Science*, **279**, 2060–206.

Stockwell, D. R. B. & Noble, I. R. (1992). Induction of sets of rules from animal distribution data: a robust and informative method of data analysis. *Mathematics and Computers in Simulation*, **33**, 385–90.

Stockwell, D. R. B. & Peters, D. (1999). The GARP modelling system: problems and solutions to automated spatial prediction. *International Journal of Geographical Information Science*, **13**, 143–58.

Stork, N. E. (1988). Insect diversity: fact, fiction and speculation. *Biological Journal of the Linnaean Society*, **35**, 321–37.

Stork, N. E. & Gaston, K. (1990). Counting species one by one. *New Scientist*, **127**, 31–5.

Sullivan, M. & Chesson, J. (1993). *The use of surrogate measurements for determining patterns of species distribution and abundance. Resource Assessment Commission Research Paper No. 8.* Canberra: Australian Government Publishing Service.

Syrjala, S. E. (1996). A statistical test for a difference between the spatial distributions of two populations. *Ecology*, **77**, 75–80.

Tabarelli, M. & Gascon, C. (2005). Lessons from fragmentation research: improving management and policy guidelines for biodiversity conservation. *Conservation Biology*, **19**, 734–9.

Takacs, D. (1996). *The Idea of Biodiversity: Philosophies of Paradise.* Baltimore: Johns Hopkins Press.

Taylor, B. L., Wade, P. R., Stehn, R. A. & Cochrane, J. F. (1996). A Bayesian approach to classification criteria for spectacled eiders. *Ecological Applications*, **6**, 1077–89.

Taylor, P. D. (2004). *Extinctions in the History of Life.* Cambridge: Cambridge University Press.

Thackway, R. & Cresswell, I. D. (eds.) (1995). *An Interim Biogeographic Regionalisation for Australia*, Canberra: Australian Nature Conservation Agency.

Thibault, M. H. (1995). *Le Québec statistique–1995.* Ste.-Foy: Bureau de la statistique du Québec.

Tognelli, M. (2005). Assessing the utility of surrogate groups for the conservation of South American terrestrial mammals. *Biological Conservation*, **121**, 409–17.

Tracy, C. R. & Brussard, P. F. (1994). Preserving biodiversity: species in landscapes. *Ecological Applications*, **4**, 205–7.

Tsuji, N. & Tsubaki, Y. (2004). Three new algorithms to calculate the irreplaceability index for presence/absence data. *Biological Conservation*, **119**, 487–94.

Tufte, E. R. (1990). *Envisioning Information.* Cheshire, Connecticut: Graphics Press.

USG Survey, (1998). GTOPO30 Global 30 Arc-second Digital Elevation Model. http://edcdaac.usgs.gov/gtopo30/gtopo30.html.

Usher, M. B. (1986). Wildlife conservation evaluation: attributes, criteria and values. In *Wildlife Conservation Evaluation*, ed. M. B. Usher. London: Chapman & Hall, pp. 3–44.

Usher, M. B. (1993). Primary succession on land: community development and wildlife conservation. In *Primary Succession on Land*, ed. J. Miles & D. W. H. Watson. Oxford: Blackwell, pp. 283–93.

van Jaarsveld, A. S., Freitag, S., Chown, S. L. *et al.* (1998). Biodiversity assessment and conservation strategies. *Science*, **279**, 2106–8.

Vane-Wright, R. I., Humphries, C. J. & Williams, P. H. (1991). What to protect? Systematics and the agony of choice. *Biological Conservation*, **55**, 235–54.

Vermeulen, S. & Koziell, I. (2002). *Integrating Global and Local Values: A Review of Biodiversity Assessment*. London: International Institute for Environment and Development.

Vijayan, V. S. (1987). *Keoladeo National Park Ecology Study*, Bombay: Bombay Natural History Society.

Villa, F., Tunesi, L. & Agardy, T. (2002). Zoning marine protected areas through spatial multiple-criteria analysis: the case of the Asinara Island National Marine Reserve of Italy. *Conservation Biology*, **16**, 515–26.

Virolainen, K. M., Ahlroth, P., Hyvarinenhj, E. *et al.* (2000). Hot spots, indicator taxa, complementarity and optimal networks of Taiga. *Proceedings of the Royal Society London Series B*, **267**, 1143–7.

von Winterfeldt, Detlof with Smith, Decision analysis in management science (2004). *Management Science*, **50**(5), 561–74.

Wade, P. R. (1994). Abundance and population dynamics of two eastern tropical Pacific dolphins, *Stenella attenuata* and *Stenella attenuata orientalis*. Unpublished Ph.D. thesis, University of California, San Diego.

Wade, P. R. (2000). Bayesian methods in conservation biology. *Conservation Biology*, **14**, 1308–16.

Wade, P. R. (2002). Bayesian population viability analysis. In *Population Viability Analysis*, ed. S. R. Beissinger & D. R. McCullough. Chicago: University of Chicago Press, pp. 213–38.

Wake, D. B. (2001). Speciation in the round. *Nature*, **409**, 299–300.

Walker, B. H. (1992). Biodiversity and ecological redundancy. *Conservation Biology*, **6**, 18–23.

Walker, P. A. & Cocks, K. D. (1991). HABITAT: a procedure for modelling a disjoint environmental envelope for a plant or animal species. *Global Ecology and Biogeography Letters*, **1**, 108–18.

Ward, T. J., Kenchington., R. A., Faith, D. P. & Margules, C. R. (1998). *Marine Bio-Rap Guidelines: Rapid Assessment of Marine Biodiversity*. Perth, Australia: CSIRO.

Weitzman, M. L. (1993). What to preserve? An application of diversity theory to crane conservation. *Quarterly Journal of Economics*, **108**, 157–83.

Wessels, K. J., Freitag, S. & van Jaarsveld, A. S. (1999). The use of land facets as biodiversity surrogates during reserve selection at a local scale. *Biological Conservation*, **89**, 21–38.

White, E., Tucker, N., Meyers, N. & Wilson, J. (2004). Seed dispersal to revegetated isolated rainforest patches in north Queensland. *Forest Ecology and Management*, **192**, 409–26.

White, G. C. & Burnham, K. P. (1999). Program MARK: Survival estimation from populations of marked animals. *Bird Study* **46**(Suppl.), 120–38.

Whitehead, P. (1990). Systematics: an endangered species. *Systematic Zoology*, **39**, 179–84.

Whittaker, R. H. (1954). Vegetation of the Great Smoky Mountains. *Ecological Monographs*, **26**, 1–80.

Whittaker, R. H. (1960). Vegetation of the Siskiyou Mountains, Oregon and California. *Ecological Monographs*, **30**, 279–338.

Whittaker, R. H. (1972). Evolution and measurement of species diversity. *Taxon*, **21**, 213–51.

Williams, P. H. & Gaston, K. J. (1998). Biodiversity indicators: graphical techniques, smoothing and searching for what makes relationships work. *Ecography*, **21**, 551–60.

Williams, P. H., Gaston, K. J. & Humphreys, C. J. (1994). Do conservation biologists and molecular biologists value differences between organisms in the same way? *Biodiversity Letters*, **2**, 67–8.

Williams, P., Gibbons, D., Margules, C. *et al.* (1996). A comparison of richness hotspots, rarity hotspots and complementary areas for conserving diversity using British birds. *Conservation Biology*, **10**, 155–74.

Williams, P. H., Margules, C. R. & Hilbert, D. W. (2002). Data requirements and data sources for biodiversity priority area selection. *Journal of Biosciences* **27** (Suppl. 2), 327–38.

Wilson, E. O. & Peters, F. M. (eds). (1988). *Biodiversity*, Washington D.C.: National Academy Press.

Wilson, K., Pressey, R. L., Newton, A. *et al.* (2005). Measuring and incorporating vulnerability into conservation planning. *Environmental Management*, **35**, 527–43.

Woinarski, J. C. Z., Whitehead, P. J., Bowman, D. M. J. S. & Russell-Smith, J. (1992). Conservation of mobile species in a variable environment: the problem of reserve design in the Northern Territory, Australia. *Global Ecology and Biogeography Letters*, **2**, 1–10.

Woodroffe, R. & Ginsburg, J. (1997). Country-by-country action plans for wild dog conservation. In *The African Wild Dog – Status, Survey and Conservation Action Plan*, ed. R. Woodroffe, J. Ginsburg & D. W. MacDonald. Gland: IUCN, pp. 118–23.

WWF/IUCN (1996). *Forests for Life – the WWF/IUCN Forest Policy Book*. Gland, Switzerland: World Conservation Union, pp. 224.

WWF-UQCN (World Wildlife Fund, Bureau de Québec – Union québécoise pour la conservation de la nature) (1998). *Les Milieux Naturels du Québec Méridional: Première Approximation*. Montréal and Québec: WWF-UQCN.

Index

Printed in the United States
by Baker & Taylor Publisher Services